SpringerBriefs in Economics

Development Bank of Japan Research Series

This series is characterized by the close academic cohesion of financial economics, environmental economics, and accounting, which are the three major fields of research of the Research Institute of Capital Formation (RICF) at the Development Bank of Japan (DBJ). Readers can acquaint themselves with how a financial intermediary efficiently restructuring firms in financial distress, can contribute to economic development.

The aforementioned three research fields are closely connected with one another in the following ways. DBJ has already developed several corporation-rating methods, including the environmental rating by which DBJ decides whether or not to make concessions to the candidate firm. To evaluate the relevance of this rating, research, which deploys not only financial economics but also environmental economics, is necessary.

The accounting section intensively studies the structure of IFRS and Integrated Reporting to predict their effects on Japanese corporate governance. Although the discipline of accounting is usually isolated from financial economics, structural and reliable prediction is never achieved without sufficient and integrated knowledge in both fields.

Finally, the environmental economics section is linked to the accounting section in the following manner. To establish green accounting (environmental accounting), it is indispensable to explore what the crucial factors for the preservation of environment (e.g. emission control) are. RICF is well-equipped to address the acute necessity for discourse among researchers who belong to these three different fields.

More information about this series at http://www.springer.com/series/13542

Katsuhisa Uchiyama

Environmental Kuznets Curve Hypothesis and Carbon Dioxide Emissions

 Springer

Katsuhisa Uchiyama
Development Bank of Japan
Tokyo
Japan

ISSN 2191-5504   ISSN 2191-5512 (electronic)
SpringerBriefs in Economics
ISSN 2367-0967   ISSN 2367-0975 (electronic)
Development Bank of Japan Research Series
ISBN 978-4-431-55919-1   ISBN 978-4-431-55921-4   (eBook)
DOI 10.1007/978-4-431-55921-4

Library of Congress Control Number: 2016937356

Printed on acid-free paper

This Springer imprint is published by Springer Nature
The registered company is Springer Japan KK

Acknowledgements

I would like to thank Yuko Hosoda very much for her various forms of assistance. I would also like to acknowledge with gratitude the valuable comments and suggestions made by the participants of the review seminar for Springer-Briefs in Economics: Development Bank of Japan Research Series, which was held on September 25, 2015. Any remaining errors, of course, are those of the author; additionally, the views expressed in this paper are those of the author, and do not necessarily reflect the official views of the Development Bank of Japan.

Acknowledgements

Contents

1 Introduction 1
  1.1 Economic Growth and the Environment 1
  1.2 Efficiency, Equity, and Global Warming 3
     1.2.1 Efficiency 3
     1.2.2 Intra-generational Equity 4
     1.2.3 Intergenerational Equity 6
  1.3 Features and Organization of the Brief 8
  References 9

2 Environmental Kuznets Curve Hypothesis 11
  2.1 Environmental Kuznets Curve Hypothesis 11
  2.2 Literature Review (1): Empirical Research 15
  2.3 Literature Review (2): Theoretical Research 20
  2.4 The Model with Stock Pollutants 21
  2.5 Problems with the Environmental Kuznets Curve 24
     2.5.1 Conceptual and Theoretical Aspects 24
     2.5.2 Empirical Aspects 26
  References 28

3 Empirical Analysis of the Environmental Kuznets Curve 31
  3.1 Data Sources 31
  3.2 Stationarity of Panel Data 34
     3.2.1 Panel Unit Roots Tests 34
     3.2.2 Panel Cointegration Tests 35
  3.3 Estimation of the Environmental Kuznets Curve 37
     3.3.1 Estimation Methods 37
     3.3.2 Estimation (1): Static Panel Data Model 38
     3.3.3 Estimation (2): Dynamic Panel Data
         (Arellano and Bond) Model 40
  3.4 Some Comments on the Turning Point 43
  References 45

4 Concluding Remarks   47
    4.1 Summary of the Study   47
    4.2 Further Research   48
    4.3 Toward the Stationary State of John Stuart Mill   49
    References   50

Appendix A: Stokey Model with Accumulative Pollutant.   51

Appendix B: Nonstationary Panel Data   55

Appendix C: Model Specification Test   59

Further Reading   61

Chapter 1
Introduction

Abstract This chapter describes the awareness surrounding the issues discussed in this Brief. The primary cause of environmental destruction is economic activity; global warming, one such type of destruction, is no exception, and its chief cause is the mass consumption of fossil fuels. The concept of sustainable development implies that the environment and economic development are not trade-offs, but rather coexist; this concept presently serves as the basis of environmental policies in every country worldwide. To attain sustainable development, it is essential to consider efficiency and equity. As for global warming, the United Nations Framework Convention on Climate Change and the Kyoto Protocol exist as central frameworks of international cooperation, with the goal of creating sustainable development. The author reviews within these frameworks efficiency and both intra-generational and intergenerational equity, and attempts to suggest new frameworks for after 2020—frameworks that are currently and frequently being discussed.

Keywords Sustainable development · Environment · Global warming · Efficiency · Equity

1.1 Economic Growth and the Environment

Since the end of the 20th century, every corner of the earth has been hit by various types of extreme weather, and the relationship between this weather and global warming has been highlighted as one of the key factors. Environmental issues have been classified into two categories: one pertains to conventional local problems that relate to environmental pollution, and the other to global problems that relate to global warming or ozone depletion. At present, there seems to be consensus that there is a close relationship between such environmental problems and economic activity, with the primary cause of environmental destruction being economic activity.

Global warming, as one form of environmental destruction, is no exception, as it derives mainly from the mass consumption of fossil fuels. Global warming is thought to have begun gradually after the industrial revolution in the mid-18th century, when humans started to consume fossil fuels excessively. One can say that we

© Development Bank of Japan 2016   1
K. Uchiyama, *Environmental Kuznets Curve Hypothesis and Carbon Dioxide Emissions*, Development Bank of Japan Research Series,
DOI 10.1007/978-4-431-55921-4_1

have enjoyed economic development and rich living standards while sacrificing the environment to global warming. According to the fifth assessment report (AR5) of the Intergovernmental Panel on Climate Change (IPCC)—which contains the most recent scientific knowledge on this matter, as of the time of writing—"Warming of the climate system is unequivocal, and since the 1950s, many of the observed changes are unprecedented over decades to millennia".[1] One can reconfirm the progress of global warming through the simple observation of such measurable parameters as atmospheric temperature, ocean temperature, sea level, and snow and ice reduction. Furthermore, AR5 also says that "It is extremely likely that human influence has been the dominant cause of the observed warming since the mid-20th century".[2] The word "likely" was used in the fourth assessment report (AR4), and so one could say that the IPCC's strength of conviction on this matter has indeed grown.

The Limits to Growth by the Club of Rome [3] is well known as a report that considers the relationship between environmental issues and economic growth and development. This report concludes that "If the present growth trends in world population, industrialization, pollution, food production, and resource depletion continue unchanged, the limits to growth on this planet will be reached sometime within the next one hundred years".[3] Additionally, the report warns of the tragedy inherent in the pursuance of economic growth on the premise of the earth's infinite capacity. Although various views have been presented concerning this conclusion, many are critical reactions, overall. For example, one of the criticisms was that the report's analyses were nothing more than extensions of a past trend. Another criticism was that the authors did not consider changes in the market prices of resources. Although 40 or more years have passed since this report was published, the crises predicted therein have not come to pass, and it seems that there are few signs. Regarding these points of concern in the report, some researchers evaluate that environmental improvements are actually progressing. Given this perspective, one might say that there is no trade-off between economic growth and the environment. The contributions of *The Limits to Growth* are that it reminds us of the significance of the fact that the environment and resource issues have been considerable constraints to economic growth, and that this fact has affected subsequent research in various ways.

The concept of "sustainable development" appeared, rather than that of "the limits to growth". At present, sustainable development is a concept widely rooted in society worldwide. The most famous and frequently quoted definition of sustainable development is that stated in Our Common Future, published in 1987 as a report of the World Commission on Environment and Development [4]. Its definition is "Sustainable development is development that meets the needs of the present without compromising the ability of future generations to meet their own needs".[4] This concept considers development and the environment as two coexisting factors, and not trade-offs; it derives from the view that moderate development in consideration

---

[1] IPCC [1], p. 4.

[2] IPCC [1], p. 17; italics original to text.

[3] Meadows et al. [3], p. 23.

[4] World Commission on Environment and Development [4], p. 43.

of environmental preservation is important. This was adopted as the basic philosophy of the "United Nations Conference on Environment and Development" (or "Earth Summit") in 1992, and it has served as a defining standard of value when considering economic development in tandem with the environment.

Although plenty of arguments have concerned the definition and meaning of sustainable development, the author does not discuss them here, as they are beyond the aim of this Brief. However, it is possible to pinpoint in various arguments a few elements that are so-called common denominators. Two are the concepts of efficiency and equity. The concept of equity, in particular, is considered an important element of sustainability, since sustainability can be considered an aspect of the problem of distribution among generations.

1.2 Efficiency, Equity, and Global Warming

On the issue of global warming, the amount of greenhouse gas emissions and the damage derived therefrom differ from country to country. To resolve this problem, it is necessary to advance measures to which each country, under international cooperation, will agree. As for the main frameworks of international cooperation against global warming, there is the United Nations Framework Convention on Climate Change, adopted at the Earth Summit in 1992. This section surveys how efficiency and equity are treated within that framework, which aims to bring about sustainable development.

1.2.1 Efficiency

Briefly, efficiency means "realizing the given target by the least cost"; indeed, it is an important concept whenever one evaluates an economy. To pursue efficiency, it becomes essential to operate a market mechanism effectively. On the issue of global warming, making the distinction between assailants and victims is difficult, and with respect to the environment, economic measures—such as tax and emission-trading— are being adopted as policies, in place of direct regulations that have frequently been used as conventional antipollution measures. Economic measures, so to speak, change the behavior of both assailants and victims by offering economic incentives; those measures look to create a sustainable society that features a healthy and viable environment.

The third article of the United Nations Framework Convention on Climate Change is entitled "Principle", and the third clause is called the "Precautionary Principle". The article requires that policies and measures associated with climate change should also consider cost-effectiveness, to generate the maximum effect at minimum expense. However, the United Nations Framework Convention on Climate Change is merely a framework, after all, and concrete economic measures have been left to rules such as those in the Kyoto Protocol.

The Kyoto Protocol, for example, introduced the Kyoto Mechanisms as economic instruments by which to maximize the net benefit of economies. Its theoretical basis is static efficiency—that is, it seems that the Kyoto Mechanisms pursue optimal emissions by minimizing the abatement costs of emissions at any point in time. On the other hand, there is also the theoretical basis of dynamic efficiency. While a static model considers an economy at a point in time, a dynamic model requires and makes use of a time axis—that is, a dynamic efficiency model minimizes the abatement cost of emissions over time, and searches for the optimal path towards a target emission level. The global warming issue is a problem that involves a very long timeframe, and so it seems that a consideration of dynamic efficiency should be important to determining the optimal path to sustainable development. However, there have been many cases in international negotiation meetings regarding global warming where short-term issues were discussed. In the Kyoto Mechanisms, for example, the commitment period was only five years (2008–2012). The second commitment period is 2013–2020. Thus, in making actual rules, it is difficult to say whether dynamic efficiency is explicitly taken into consideration. Such measures may in fact be detrimental to the efficiency of economies.

1.2.2   Intra-generational Equity

When discussing sustainable development, a consideration of efficiency is of course necessary, but equity is a much more important issue.

Under a market economy system, distribution is also determined by the market mechanism. For instance, in terms of income distribution, persons with scarcity and high market valuation tend to receive a high income, while persons without them may, in extreme cases, receive no income at all. Under a market economy system, this is a very natural state, and unless the government or a third party tries in some way to perform income distribution in a compulsory fashion, the efficiency of resource allocation by way of the market mechanism will become impaired. Obviously, this is not desirable.

In the case of environmental problems and a market mechanism, it is sometimes necessary to introduce private property rights—like the environmental right of use—and assign them appropriately. An emission-trading system emulates such a structure. A market mechanism with private property rights does achieve efficient resource allocation, but always ignores equity of distribution. Sustainable development is considered a distribution problem among generations, regarding the costs and benefits that stem from environmental problems. However, one cannot expect equity of distribution to be realized solely through the pursuit of efficiency by way of a market mechanism.

As for equity, one can classify it into two categories: there is intra-generational equity (e.g., equity among countries) and intergenerational equity. The author considers intra-generational equity.

Considering equity among countries, it is difficult to reach consensus in the absence of an institutional framework of international cooperation—a framework that would contain the substance of equity among countries, since, as a feature of global warming, the amount of greenhouse gases emissions and damage from global warming differ from country to country. However, in the present condition—in which a super-national organization that carries out alliances does not exist—it is very difficult to build an institutional framework. Confrontation based on the viewpoint of equity has actually been seen, mainly between advanced countries and developing countries, during negotiations in the United Nations Framework Convention on Climate Change. The points of conflict were as follows.

The first point concerns the argument regarding the responsibilities of advanced countries. Advanced countries assert that global warming is a problem that the whole world faces, and that each country should take common responsibility in line with the amount of greenhouse gases it discharges. On the other hand, developing countries claim that they have the "right to development", and that advanced countries should also help pay the costs associated with measures, as they too had contributed to global warming in the process of their economic development. As a result of these assertions, the following sentences were added to the third article, first clause of the framework: "The Parties should protect the climate system for the benefit of present and future generations of humankind, on the basis of equity and in accordance with their common but differentiated responsibilities and respective capabilities. Accordingly, the developed country parties should take the lead in combating climate change and the adverse effects thereof". This is well known as the principle of "common but differentiated responsibilities". Then, considerations for equity—which infringes upon neither the responsibility of advanced countries nor the opportunity for development among developing countries—were incorporated into the international cooperation framework.

The second point pertains to what arises from the difference in coping capacities between advanced and developing countries. In particular, the capacities relate to the establishment of greenhouse gas emission targets, and the allocation of total worldwide emissions to each country. Since numerical targets can also serve as index values that relate to the responsibility of each country, they can also relate to the aforementioned common but differentiated responsibilities. While advanced countries (save for the United States) have approved of target-setting, the United States has shown an unwillingness to set targets, because there is no scientific basis for the target, and the measures incur enormous costs. The opinions of developing countries are that the advanced countries are actually the polluters, and that developing countries can perform possible measures only with the assistance of advanced countries. After all, the framework requested that measures be drawn only for the advanced countries, and the targets were set to 1990 levels.

The third point pertains to financial support. It is the problem of desirable cost-sharing with regards to global warming measures, between advanced and developing countries. Advanced countries have insisted that they install funds in the Global Environment Facility (GEF), which was established mainly by the World Bank as an advanced form of the "polluter pays principle", and that through this funding, they will support the developing countries. On the other hand, developing countries claim that there is a need to establish a new organization for financial support, as the GEF is merely a substitute organization for the advanced countries, so to speak. After all, the framework positions GEF as a provisional financial support organization. Since a direct burden occurs with respect to financial support, at present, advanced countries do not necessarily tackle the issue of financial assistance positively.

As seen above, considerations of intra-generational equity (i.e., equity among countries) have been indispensable in the bargaining process of international cooperation against global warming. However, many people consider it a fault of the Kyoto Protocol, that developing countries do not have a duty to abate their greenhouse gas emissions. Although those countries that have a negative attitude toward international cooperation are anxious about its influence on their economic growth, the involvement of developing countries remains a key factor in resolving the issue of global warming, since it is predicted that the amount of emissions from developing countries already exceed those of advanced countries (Fig. 1.1). From the viewpoint of equity, it is desirable for the advanced countries to assume considerable responsibility with regard to abatement obligations or funding contributions. In addition, it is also desirable that institutional arrangements be designed in which all countries—including developing ones—will participate after 2020.

1.2.3   Intergenerational Equity

The issue of global warming is considered a problem of intergenerational equity over a very extended time period; at the same time, it is a problem of intra-generational equity. Suppose a distribution of the present generation on the basis of a market economy system lacks equity; in such a case, the inequities of distribution in the future generation will tend to expand throughout the generations. Therefore, the argument on intergenerational equity is inescapable with respect to the global warming issue, which is an ultra-long term problem. We need to promise to future generations the same level of economic prosperity that the present generation enjoys, and also leave the same level of environmental conditions to those future generations. Intergenerational equity relates directly to the concept of sustainable development, in that sense.

Solving an intergenerational problem that relates to global warming is very difficult, for a number of reasons. Although past and present economic activities have greatly influenced global warming, the past generation of assailants no longer exists, and the future generation—which will be victims—does not yet exist. In resolving external diseconomies and establishing appropriate policies, standard economic theory assumes local environmental degradation (e.g., pollution), with assailants and victims who are separable and exist simultaneously. Global warming is a problem that

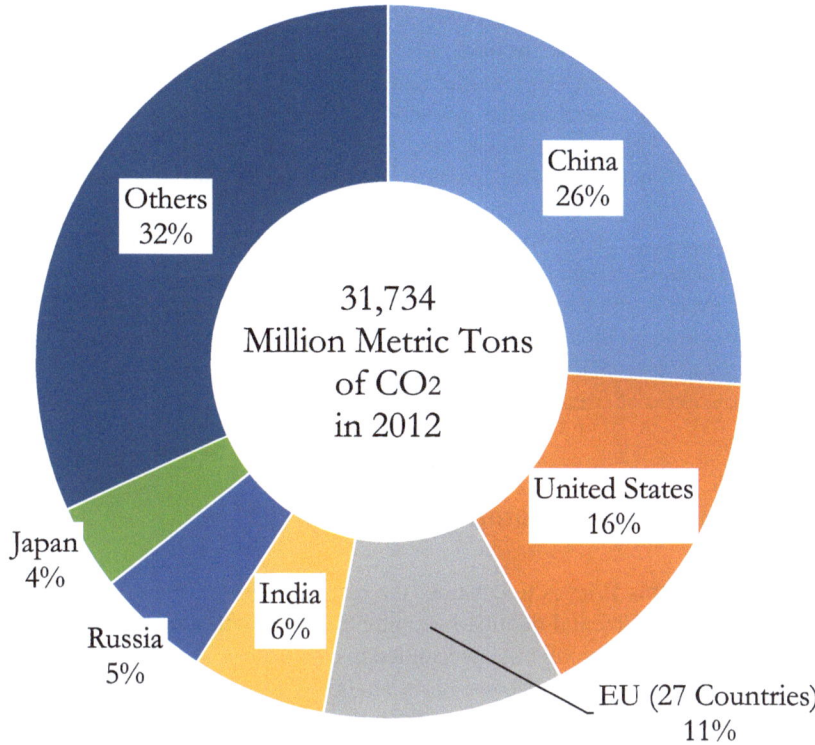

China
26%

Others
32%

31,734
Million Metric Tons
of CO2
in 2012

United States
16%

Japan
4%

India
6%

Russia
5%

EU (27 Countries)
11%

(Note) EU does not include Croatia (joined in 2013).
(Source) CO2 Emissions from Fuel Combustion (2014 Edition), IEA, Paris.

Fig. 1.1 Percentages of energy-related carbon dioxide emissions worldwide

is widespread in space and time, and it is one that human beings have not experienced until very recently. Therefore, it may be impossible for one to fully comprehend the issue through the use of the frameworks of existing economic theory. There is also the question of its application. The problem of intergenerational equity can be said to be a more fundamental problem, in that sense.

Not only the people of the present generation but also those of future generations have an equal right to use the services the atmosphere provides. However, it is not the present generation—but rather, the future generation—that more seriously suffers from damage owing to global warming. As mentioned, although global warming measures such as the Kyoto Mechanisms take into consideration to some extent intra-generational equity, it seems that they little consider intergenerational equity. The concept of equity, even though it is intra-generational or intergenerational, assumes a value judgment. The criterion of judgment regarding what is equitable for present and future generations differs at the individual level, and so the concept contains more elements of politics, sociology, ethics, and social philosophy than it does economics. Therefore, one can say it is very difficult to handle the concept of equity.

There are some points of consideration that arise when discussing intergenerational equity. The first is the discount rate. While the cost of the measures that counter global warming is a short-term matter and the present generation will mainly cover it, the benefits that arise from the measure constitute a long-term matter and future generations will enjoy them. Therefore, if the discount rate is high, future benefits will be underestimated; such measures might not even be considered in cost-benefit analyses.[5]

The second is about decision-making. Although it is the present generation that makes decisions with regard to global warming measures, future generations should also participate in the decision-making, from the viewpoint of equity. Obviously, this is impossible. Although the present generation can act as a representative of the future generation—which cannot participate in the decision-making—this cannot satisfy the criteria of equity.

1.3 Features and Organization of the Brief

The purpose of this Brief is to consider the relationships between economic development and environmental quality, by estimating the environmental Kuznets curve (EKC). The global warming issue is handled in this Brief as a specific environmental problem, for the following reasons: (1) when considering sustainable development, it is necessary to examine the problem from the vantage of a very long time period; therefore, global warming is thought to be a typical subject; (2) many EKC research papers that touch upon carbon dioxide (CO_2) emissions have been published thus far, with various controversial results regarding the existence of the EKC; therefore, the EKC for CO_2 remains a matter of further research; and (3) various international measures have been undertaken to address global warming, and so researchers are interested in whether or not some structural changes with regard to CO_2 emissions have arisen on account of such measures.

The features of this Brief are as follows. First, it estimates the EKC with regard to CO_2. Many previous studies on EKC estimation have employed local air pollutants such as sulfur dioxide (SO_2) and nitrogen oxide (NO_X) as part of a pollution index. On the other hand, there have not been many studies concerning EKC estimation for CO_2, and so researchers have not yet reached a clear conclusion on EKC formation. Second, the Brief adopts sophisticated methods of EKC estimation—that is, data stationarity tests, and a dynamic panel data estimation model. Third, the Brief utilizes the most recent available data, to 2010. Fourth, this Brief is interested in the income levels inherent in the EKC turning point for a sustainable society, while many other studies have mainly focused on the shape of the EKC, or EKC estimation techniques.

This Brief is organized as follows. Chapter 2 provides a brief overview of the EKC hypothesis, which is used to examine relationships between economic development

---

[5]See Kuninori and Otaki [2] for a discussion of discount rate.

and the environment; this chapter also surveys the EKC literature, from theoretical and empirical perspectives. Furthermore, it explains some problems highlighted in previous studies. Chapter 3 performs empirical analyses to estimate EKC with regard to CO_2 emissions, while considering the problems raised in previous studies. Chapter 4 provides concluding remarks.

References

1. IPCC (2013). *Climate change 2013: The physical science basis (Summary for policymakers)*, Intergovernmental panel on climate change. http://www.climatechange2013.org/images/report/WG1AR5_SPM_FINAL.pdf.
2. Kuninori, M., & Otaki, M. (2015). The revelation of the time preference rate and intertemporal negative externality, RCGW Discussion Paper Series 53, Development Bank of Japan.
3. Meadows, D. H., Meadows, D. L., Randers, J., & Behrens, W. (1972). *The limits to growth*. New York: Universe Books.
4. World Commission on Environment and Development (1987). *Our common future*. Oxford: Oxford University Press.

Chapter 2
Environmental Kuznets Curve Hypothesis

Abstract This chapter deals with the environmental Kuznets curve (EKC) hypothesis, as one hypothesis that speaks to the possibility of sustainable development; this chapter also provides a brief overview and current research trends. The EKC hypothesis implies there is an inverted U-shaped relationship between environment degradation and economic development. The concept of the EKC originally began with data observation, and has attracted considerable attention from many researchers concerning the existence of the EKC—since it is very naive—and the thinking that relates to it. This chapter also surveys the main EKC research, which has been undertaken from both empirical and theoretical perspectives. As for the empirical perspective, the author compares previous studies that focus on the income level inherent in the EKC turning point. From the theoretical perspective, the author briefly introduces the Stokey (International Economic Review, 39, 1–31, [43]) model, which addresses stock pollutants. Finally, the author takes up criticisms and problems inherent in previous empirical and theoretical studies of the EKC.

Keywords Environmental Kuznets curve · Inverted U-shaped · Turning point · Carbon dioxide · Income level

2.1 Environmental Kuznets Curve Hypothesis

When considering sustainable development that will take place in the future, it is important to accurately grasp past circumstances and the present situation regarding the relationship between economic development and environmental quality. This section surveys the empirical and theoretical studies of the environmental Kuznets curve (EKC) hypothesis and points out the problems therein.

In the field of environmental economics, the EKC hypothesis is used to consider the relationship between economic growth and the environment. This hypothesis was derived through data observation. When one plots per-capita income along a horizontal axis and the per-capita index of environmental degradation on a vertical axis for a certain country, he or she will generally find a relationship that takes the form of an inverted U-shaped curve. In other words, environmental degradation increases

© Development Bank of Japan 2016

K. Uchiyama, *Environmental Kuznets Curve Hypothesis and Carbon Dioxide Emissions*, Development Bank of Japan Research Series, DOI 10.1007/978-4-431-55921-4_2

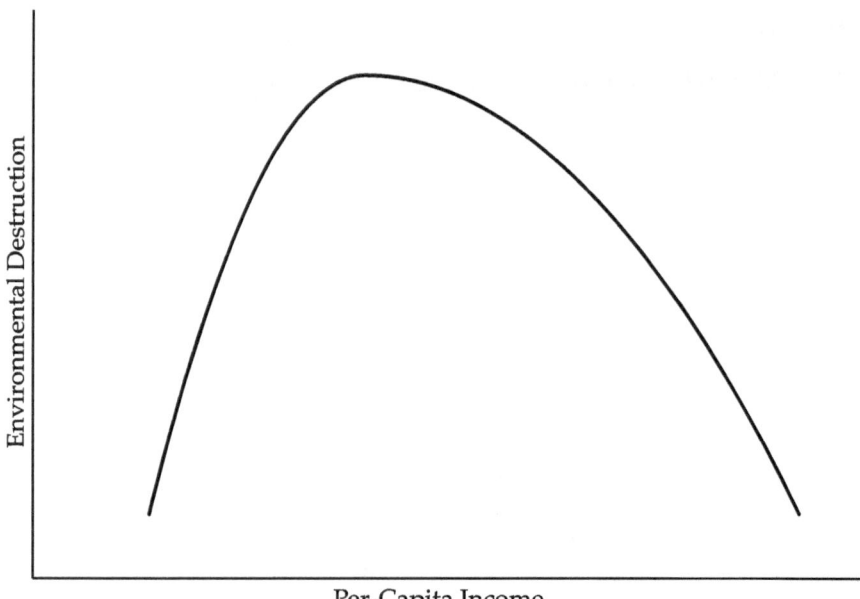

Fig. 2.1 Environmental Kuznets curve

in the early stage of economic development, and after per-capita income exceeds a certain level (i.e., turning point), it turns to decrease as income increases (Fig. 2.1). Simon Kuznets found the relationship between per-capita national income and income inequality, and he went on to present the hypothesis that the relationship grows in the early stage of economic development and shrinks in later stages [26]. This is known as the Kuznets curve hypothesis, and the EKC is an analogy of it. Recently, when considering sustainable development, it has been more important to understand the influence that economic growth has on the environment, and so the concept of the EKC has also become attractive when pursuing sustainable development.

Economic research on the EKC began with Grossman and Krueger [21] and Shafik and Bandyopadhyay [39]; the latter of these served as the background study of the report of the World Bank [47]. The concept then came to be known widely, by virtue of the World Bank. Although these studies made use of data captured through observation, their findings came to be considered pieces of evidence that suggest the feasibility of sustainable development, provided the existence of the EKC becomes clear through analysis. Moreover, there was eventually the expectation that the EKC could help predict long-term environmental changes and thus assist in crafting preventive policies. Comprehensive survey studies in this field include those of Stern [40, 41], Panayotou [32], Dasgupta et al. [11], and Dinda [14], inter alia. In

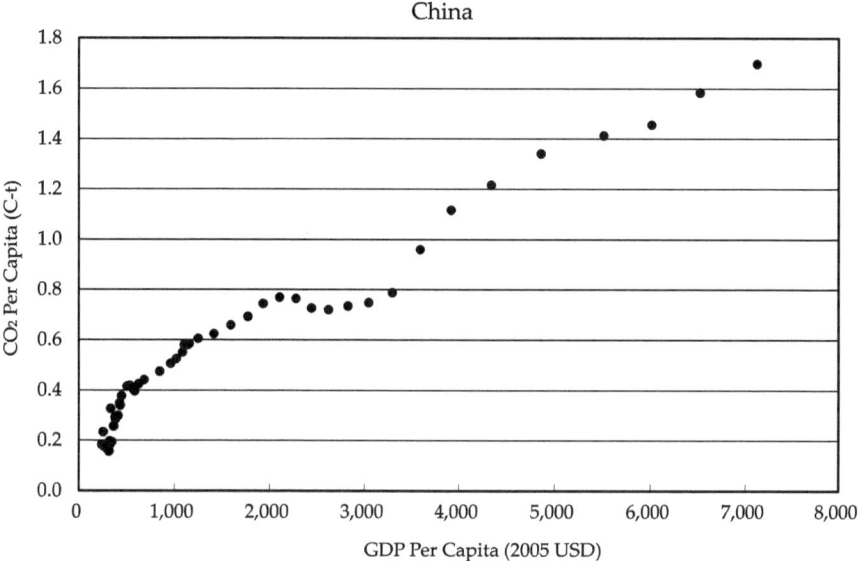

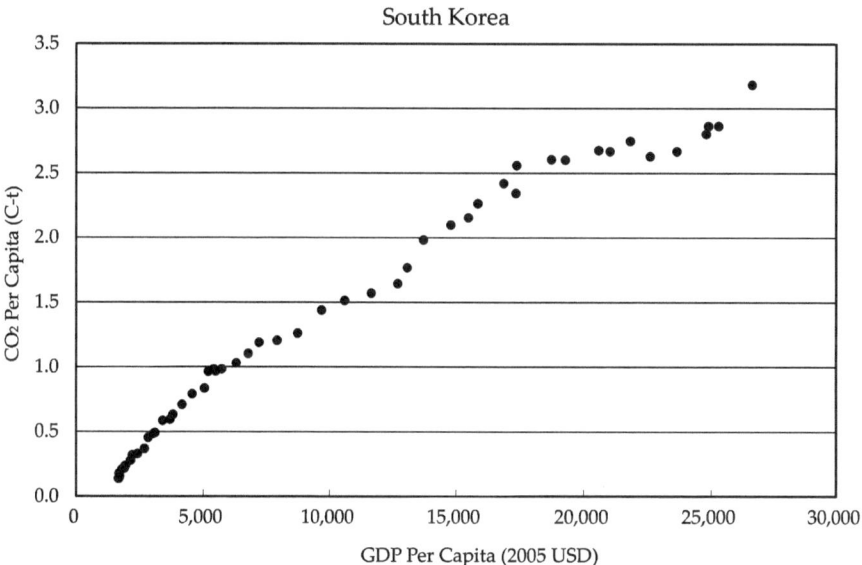

Fig. 2.2   Environmental Kuznets curve of each country (developing countries)

the late 1990s, some special issues of academic journals were dedicated to the topic of the EKC[1] and attracted considerable research interest.

---

[1]For instance, see *Environment and Development Economics*, Vol. 2, Part 4, 1997, and *Ecological Economics*, Vol. 25, No. 2, 1998.

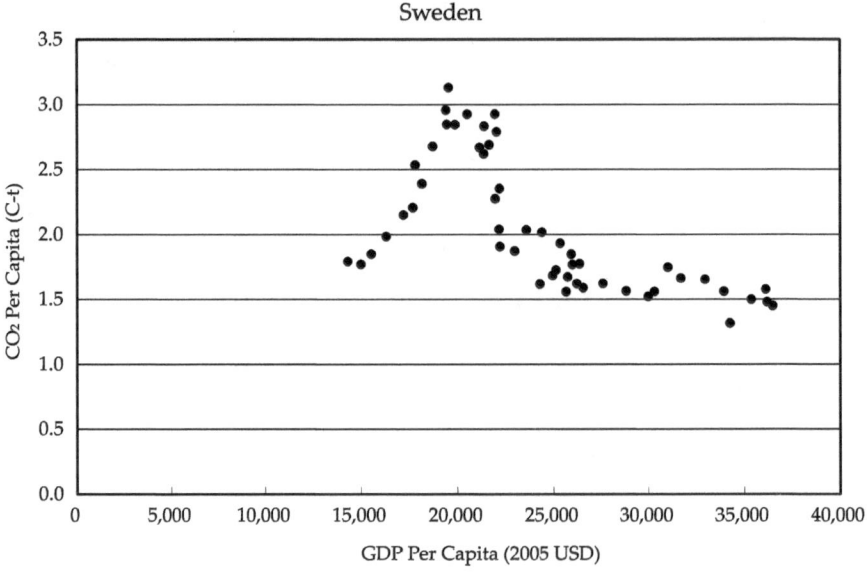

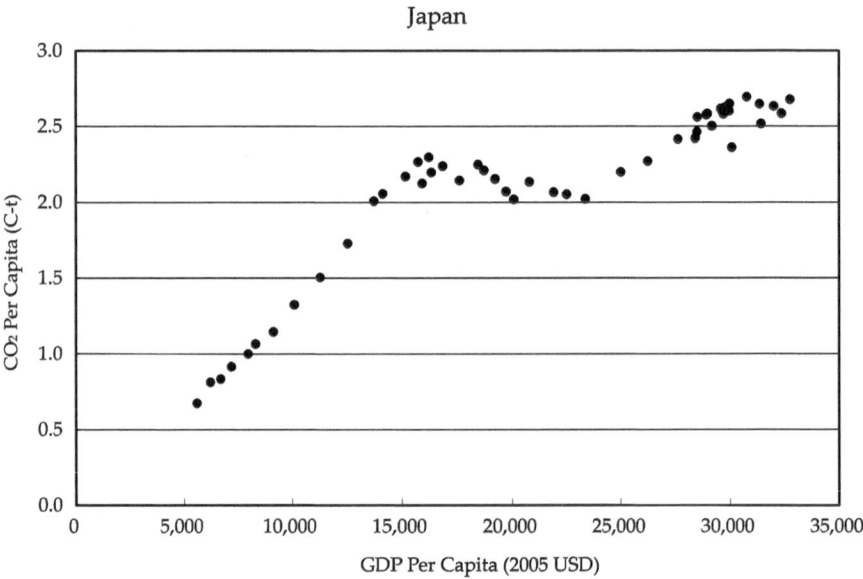

Fig. 2.3 Environmental Kuznets curve of each country (advanced countries)

Typical EKC patterns are shown in Fig. 2.1. The pattern for each country becomes part of the inverted U-shaped curve naturally, in line with that country's stage of economic development. The author takes up the cases of Sweden, Japan, South Korea, and China. Figures 2.2 and 2.3 plot the relationship between per-capita carbon dioxide (CO_2) emissions and the real GDP of the countries. China, a developing country, generally shows an upward-sloping curve. Although it shows a downward slant just after the Asian currency crisis, it returned to an upward slant more recently, and the rate of increase in CO_2 emissions per GDP has also increased. South Korea is treated as a developing country in this study, as it has been classified as part of the non-Annex I group of the Framework Convention on Climate Change—although South Korea could be considered an advanced nation, given its OECD membership. Looking at the figure of South Korea, one can find a consistent upward slanting curve, as if South Korea were a developing country. Although a fall in CO_2 emissions in both South Korea and China was seen at the time of the Asian currency crisis, the figure shows an upward rise in South Korea recently, and it has not yet reached the turning point. Sweden, an advanced nation in northern Europe that is seen as very environmentally conscious, reached its turning point after an upward sloping curve; it seems to now be in the downward-sloping stage. Sweden's CO_2 emissions are decreasing, concurrent with an increase in income, and the full EKC is obvious. Although Japan is also an advanced nation, its figure differs from that of Sweden. CO_2 emissions in Japan increased during the period of high economic growth, as income increased. The amount of emissions came to decrease along with an increase in income, probably because energy-saving progressed after the two oil crises—that is, the EKC of Japan seemed to appear, along with a turning point. However, the emissions increased again after the bubble economy period, when the EKC took on the appearance of one from a developing country. As a result, it takes on more of an N-shaped curve than an inverted U-shaped curve. Not only income but also the industrial structure of each of economy, energy composition, trade, and others is considered a factor that brings about this different shape. Analysis has also been directed toward such factors.

2.2 Literature Review (1): Empirical Research

To date, many studies have undertaken empirical analyses of the EKC. There are some reasons for this research interest. The studies in this field began with empirical analyses, and researchers are interested in the EKC because the underlying concept is naive and imperfect and has many unresolved points.[2] Since it is difficult to cover such a large body of work in its entirety, the author will introduce some of the more typical studies.

In general, empirical studies have sought to explain per-capita environmental degradation level by way of a polynomial equation of per-capita income, ever since the pioneering research of Grossman and Krueger [21].[3] A standard regression model

---

[2] As mentioned in Sect. 2.5, many problems have been pointed out with regard to the EKC.

[3] Some empirical studies employ per-income CO_2 emissions as a dependent variable. However, this is an index of energy efficiency, and so it may be inappropriate as an index of environmental degradation level.

is as follows, and it has recently been used with panel data in many empirical studies.

$$E_{it} = \beta_0 + \beta_1 y_{it} + \beta_2 y_{it}^2 + \beta_3 y_{it}^3 + \beta_4 x_{it} + u_{it}, \qquad (2.1)$$

where E denotes the index of environmental pollution, y expresses per-capita income, and x is other control variables. Moreover, i denotes countries, and t is time.

If $\beta_1 > 0$, $\beta_2 < 0$, and $\beta_3 = 0$ with statistical significance, and the turning point where the pollution level starts to fall is within a reasonable range, the EKC is thought to exist. In the model of Eq. (2.1), the per-capita income level of the turning point is given by $(-\beta_1/(2\beta_2))$. The major concern of empirical analyses is how high the income level is that corresponds to the turning point.

When analyzing natural resources like forests, not a standard model resembling Eq. (2.1) but rather a log-linear and semi-log-type model is often used, as the dependent variable cannot be allowed to fall below zero.

In the case of $\beta_3 > 0$ with statistical significance, the decrease in the pollution level—along with an increase in income—is considered temporal; conversely, the pollution level will worsen at a stage where there is higher income. That is, this type of estimation model is based on the hypothesis that the EKC is an N-shaped curve.[4] x is introduced to control the variables that can affect both the environment and income (e.g., trade and energy prices). Agras and Chapman [1] survey the effect of explanatory variables other than income, and conclude that income has the greatest explanatory power. The results of Stern [40] and Neumayer [30] align with this result.

The techniques used in empirical studies can be classified into a few different categories. The first is the country-based category, where EKC estimates are drawn for the whole world or specific countries. From the 1990s to the 2000s, many studies focused on estimating the EKC of the whole world. These studies assume that all countries have the same locus. As one can easily imagine, however, various conditions—such as natural conditions and social situations, all of which can affect economic development—differ among countries in reality. Moreover, in developing countries, there is the considerable possibility that the curve can shift to the lower left on account of the so-called latecomer advantage. Numerous research studies that examined EKC estimations by using data from specific countries have been published; Al-Mulali et al. [2] provides a review of such studies.

The second category is that based on data characteristics and estimation methods. The per-capita income and pollution level data may be nonstationary; when data are nonstationary, regression analysis can give rise to spurious correlation, and results can be rendered meaningless. To preclude this, some studies test data stationarity, and then employ the autoregressive distributed lag (ARDL) model or error-correction model after confirming cointegration. Wagner [46] carries out panel unit root tests and panel cointegration tests for CO_2. Agras and Chapman [1] and Al-Mulali et al. [2] are studies that make use of the ARDL model. As for estimation methods, some

---

[4]Grossman and Krueger [22] and Vincent [45] each employed a cubic polynomial model while assuming an N-shaped EKC; each obtained statistically significant results. As shown in Fig. 2.3, the Japanese CO_2 data seem to illustrate an N-shaped curve.

studies use a nonparametric method—which is flexible for estimation (e.g., Ordás Criado [31])—or a calibration method (e.g., Egli and Steger [17]; Bartz and Kelly [5]).

To date, empirical studies have addressed atmospheric quality, water quality, waste, city sanitation, and energy use. Especially, many have focused mainly on air quality, such as sulfur dioxide (SO_2), nitrogen oxide (NO_X), carbon monoxide (CO), suspended particulate matter (SPM), and CO_2.

The following are typical studies on air quality, on elements other than CO_2. Selden and Song [36] conducted analyses of SO_2, SPM, NO_X, and CO, and conclude the existence of the EKC. The turning point is a level at around USD10,000 (1985 constant prices). The characteristic point is adding population density to the explanatory variables. Grossman and Krueger [22] investigated 14 kinds of environmental indicators with regard to air quality and river water quality, and confirm the formation of the EKC for almost all the environmental indicators; there, for many environmental indicators, the turning point is less than USD8,000 (1985 constant prices). Cole et al. [9] used some additional explanatory variables and support the formation of the EKC for local air pollution. Although the turning point differs with the pollutant, it was found to range from USD6,000 to USD18,000 (1985 constant prices), which is within the range of the sample. Stern and Common [42] targeted sulfur emissions and found the existence of an inverted U-shaped curve; they also show that the turning point is at a level (USD9,181; 1990 constant prices) appropriate to their sample of OECD countries, and that the turning-point level becomes extremely high with a monotonic increase for the whole world and for the sample of non-OECD countries. Thus, it is clear that sample selection bears an important influence.

As mentioned, regarding some environmental indicators of the air pollutants that have direct effects on human health, plenty of studies support the formation of the EKC. However, no consensus has been reached about the turning-point level at which environmental improvements start.

On the other hand, few studies address global pollutants such as CO_2, compared to other local air quality factors, and researchers have not reached a clear consensus about EKC formation. As for CO_2, some typical studies are as follows.[5] Shafik [38] builds upon the report of the World Bank [47] and illustrates an EKC with a monotonic increase. Holtz-Eakin and Selden [23] find an inverted U-shaped relationship, but an extremely high turning point of USD35,428 to USD8 million (1985 constant prices) or more; in other words, this is substantively equal to a monotonic increase. Schmalensee et al. [35] confirm the negative income elasticity of environmental pollution in high-income countries, and this supports the existence of the EKC with a reasonable turning point.

---

[5]Previous empirical studies investigated only CO_2 emissions; to the best of the author's knowledge, no research has been undertaken that considers the source of CO_2 absorption.

Agras and Chapman [1] adopted energy price and the trade-GDP ratio as explanatory variables, and applied the ARDL model (ARDL(1,0)) to analyze the dynamic process. Their results point out that the energy price can serve as an important explanatory variable, and that the turning point is USD13,630 (1985 constant prices). Neumayer [30] applied natural factors such as temperature and the rate of growth of a country's economic activity area to the explanatory variables. Although Cole [7] considers the influence of trade and shows the formation of the EKC, the turning point differs with the combination of explanatory variables adopted, and thus ranges from USD20,352 to USD56,696 (1985 constant prices).

Dijkgraaf and Vollebergh [13] obtained USD20,600 (1990 constant prices) as the turning point, and found that the estimated parameters of each country are not homogeneous. This means that there is no EKC common to each country, and they insist that it is necessary to allow heterogeneity in the income parameters and in the individual effects of each country. Richmond and Kaufmann [34] adopted the share of various kinds of fuels as explanatory variables, to preclude variable lack bias. Although they confirmed the formation of the EKC with turning points of USD29,687 and USD25,450 (1996 constant prices) for the whole world and for OECD members, respectively, they could not show the existence of the EKC through the use of a non-OECD sample. Galeotti et al. [20] compared the estimation results by using data from the International Energy Agency (IEA) and from the Oak Ridge National Laboratory (ORNL). In addition, another characteristic of this study is the adoption of a nonlinear estimation model, which differs from a conventional polynomial equation of income. Their results support the formation of the EKC, and although the turning point differs by estimation model and data source, it ranges from USD15,698 to USD16,595 (1990 constant prices) for the sample of OECD countries, and between USD15,600 and USD21,186 for the non-OECD sample (Table 2.1).

Thus, there are two cases for the EKC, with regard to CO_2 emissions: one with an inverted-U curve, or another that is monotonically increasing. In undertaking a literature review, one can find the tendency for studies whose estimation period ends in 1995 or later to conclude on the existence of the EKC. Note, however, that most of those studies' estimated turning points are over USD20,000; this level is high, compared to those of other pollutants. Some studies point out that the EKC with regards to CO_2 is not observed for the whole world, even if it is observed for individual countries.

In summary, the EKC with regards to air quality is observed when the pollution area is local, the pollutant is of a flow-type and decomposes in a relatively short time, and some regulations have already been introduced. When, as is the case with CO_2, the pollutant is of a stock-type and few regulations have been introduced, the existence of the EKC remains controversial.

Table 2.1 Empirical studies of the environmental Kuznets curve for carbon dioxide

	Data source	Data period	Countries	Shape of the curve	Turning point	Characteristics of the model
Shafik [38]	ORNL PWT	1960–1989	153	Monotonic increase	N.A.	
Holtz-Eakin and Selden [23]	ORNL PWT	1951–1986	130	(Inverted-U)	35,428–8,000,000 (1985 USD)	Concluding monotonic increase substantively
Cole et al. [9]	ORNL PWT	1960–1991	7 regions	(Inverted-U)	25,100–62,700 (1985 USD)	Concluding monotonic increase substantively
Schmalensee et al. [35]	ORNL PWT	1950–1990	141	Inverted-U	N.A. (1985 USD)	
Agras and Chapman [1]	ORNL PWT, UN	1971–1989	34	Inverted-U	13,630 (1985 USD)	Employing trade indicators and energy price in ARDL model
Galeotti and Lanza [19]	IEA OECD	1971–1996	110	Inverted-U	15,073–16,646 (1990 USD)	Employing nonlinear model
Neumayer [30]	ORNL PWT	1960–1988	106	(Inverted-U)	N.A. (1985 USD)	Considering natural factors and concluding monotonic increase substantively
Cole [7]	ORNL PWT	1975–1995	32	Inverted-U	20,352–56,696 (1985 USD)	Considering trade
Dijkgraaf and Vollebergh [13]	OECD	1960–1997	24	(Inverted-U)	20,647 (1990 USD)	No EKC is estimated with common parameters of income for each country
Richmond and Kaufmann [34]	IEA PWT	1973–1997	36	Inverted-U	29,687 (1996 USD)	Employing the share of fuel types
Galeotti et al. [20]	IEA, ORNL OECD, WDI	1960–1997	N.A.	Inverted-U	15,600–21,186 (1990 USD)	Employing nonlinear model

Note ORNL Oak Ridge National Laboratory, *PWT* Penn World Table, *WDI* World Development Indicators

The parenthesis in the Shape of the Curve implies the external formation of the inverted-U curve, not the substantive one

2.3 Literature Review (2): Theoretical Research

As seen in Sect. 2.2, plenty of empirical studies have been actively undertaken since the emergence of the EKC concept, and they are based on various environmental indices. In that sense, the EKC can be said to encompass "stylized facts", so to speak. Although there were nothing more than a few theoretical analyses of the EKC in the early 1990s, many theoretical studies that offer consistent explanations of the stylized facts have been published since the mid-1990s, along with an increase in the need for a theoretical background.

Many theoretical models have been built and can be classified into a few different categories. The most common classification is the static model versus the dynamic model.[6] Various theoretical studies with the background of a dynamic model have been presented so far; they attempt to describe changes in environmental quality in the process of a country's economic development, over time.

John and Pecchenino [24] adopted an overlapping generations model and derived the EKC of stock-type pollution, which is discharged mainly with consumption. Selden and Song [37] applied a modified Ramsey–Cass–Koopmans model—which is based on Forster [18]—and analyzed flow-type pollution emits from production processes. They conclude that there is the possibility of an inverted U-shaped curve emerging, under some conditions. Stokey [43] introduced the AK model and the Ramsey–Cass–Koopmans model; that study assumes pollution is discharged from production processes and analyzes the pollution of both flow and stock types. That study also suggests that to improve social welfare, environmental regulations that reduce pollutant emissions should be introduced.

Lopez [28] and Bulte and van Soest [6] each address the depletion of natural resources. Dinda [15] employed the framework of an endogenous growth model to examine the stock-type environment; this is used for both production and the abatement of pollution.

Lopez [28] and Stokey [43] made use of both static models and the aforementioned dynamic models. There are other static models, as follows. Lieb [27], in an extension of the static model of McConnell [29], analyzed pollution arising from consumption, and shows that the satiation of consumption is a necessary and sufficient condition for the existence of the EKC. This condition is drawn from the assertion that the increment of income turns, to be spent on pollution abatement. Andreoni and Levinson [3] indicate that economies of scale in pollution abatement are sufficient conditions for the existence of the EKC.

Some explanations have been given for the formation of an inverted U-shaped curve. Andreoni and Levinson [3] offer the following: (1) changes of composition of production and consumption, (2) a stronger preference for the environment, (3) the introduction of institutions that internalize external diseconomies, and (4) the effects of increasing returns to scale in abatement activities. de Bruyn and Heintz [12] point out five factors: (1) behavioral change and preference, (2) institutional

---

[6]Other classifications commonly seen are macroeconomic versus microeconomic, long-term versus short-term, and deterministic versus stochastic, among others [25].

changes, (3) technological and organizational changes, (4) structural changes, and (5) international reallocation. As mentioned, various models have been used in the theoretical derivation of the EKC, but they are contingent on the assumptions or values of parameters. To clarify this point—and also to consider the relationship with the empirical analyses in Chap. 3—the author will survey a typical theoretical model.

2.4 The Model with Stock Pollutants

This section surveys the study of Stokey [43] as a typical study that derives the EKC theoretically, and within the framework of optimum growth theory.

Stokey [43] presents four kinds of models—namely, the (a) static model, (b) dynamic model with flow-type pollutants, (c) dynamic model of (b) with exogenous technological progress, and (d) dynamic model with stock-type pollutants. Global warming relates to the accumulation in the atmosphere of greenhouse gases like CO_2; therefore, the author surveys a (d)-type model, which is conscious of the stock pollutant CO_2.

In this model, the following Cobb–Douglas production function with exogenous technological progress is assumed:

$$Y(t) = Ae^{gt}K(t)^{\alpha}L^{1-\alpha}z(t), \tag{2.2}$$

where $Y(t)$ is output, A denotes the parameter of productivity, g represents the exogenous rate of technological progress,[7] $K(t)$ is the amount of capital stock, and $L(t)$ is the labor input of the economy. α is assumed to $\alpha \in (0, 1)$.

$z(t)$ is the index of environmental technology and is $z(t) \in [0, 1]$. $z(t) = 1$ expresses the production technology with the maximum pollutant emissions; conversely, $z(t) = 0$ means the cleanest production technology is employed.

Pollutants are discharged only from the production process; the author does not address cases such as the waste problem, where consumption affects the environment. The pollution emission function is:

$$x(t) = Ae^{gt}K(t)^{\alpha}L^{1-\alpha}z(t)^{\beta}, \tag{2.3}$$

---

[7]Stokey [43] adopts the AK model with an environmental constraint to models (a) and (b). According to these models, the economy converges to the steady state, in which the growth rates of all the variables are zero as a result. This is because, as shown later, the marginal productivity of capital ($\partial Y(t)/\partial K(t) = \partial(AK(t)z(t))/\partial K(t) = Az(t)$) monotonically decreases as $z(t)$ monotonically decreases over time. This means that the long-term growth rate of the economy becomes zero and sustainable development becomes infeasible. To modify this, exogenous technological progress is introduced to models (c) and (d). The exogenous technological progress brings about increased productivity, and compensates for the fall in $z(t)$; hence, the growth rate becomes positive in the steady state.

where $x(t)$ is the emission of pollution flow in each time period, β is the parameter that shows the relationship between environmental technology and pollution flow, and $\beta > 1$.

When $z(t) = 1$, the output and emission become the maximum. Stokey [43] describes the output at this time as "potential output". If $z(t)$ falls from 1, the environment improves with the introduction of cleaner technologies, but the output decreases. When $z(t) = 0$, both the output and emissions reach zero.

The per-capita output is:

$$y(t) = Ae^{gt}k(t)^{\alpha}z(t), \tag{2.4}$$

where $y(t) \equiv Y(t)/L$ denotes per-capita output, and $k(t) \equiv K(t)/L$ is the per-capita amount of capital stock. Then, the author can write Eq. (2.3) as

$$x(t) = Ae^{gt}k(t)^{\alpha}Lz(t)^{\beta}. \tag{2.5}$$

The instantaneous utility function of the representative agent with an infinite horizon at time t is

$$u(t) = \frac{c(t)^{1-\sigma} - 1}{1 - \sigma} - \frac{B}{\gamma}X(t)^{\gamma}, \tag{2.6}$$

where $u(t)$ expresses the utility of the representative agent, $c(t)$ is per-capita consumption, σ denotes the parameter of the Arrow–Pratt measure of relative risk aversion (inverse of the intertemporal elasticity of substitution of consumption), and $\sigma > 0$. B and γ are the parameters that express disutility from pollution stock, and $B > 0$ and $\gamma > 1$. $X(t)$ shows the level of pollution stock in the economy.

The first term on the right-hand side of Eq. (2.6) is a standard constant relative risk aversion (CRRA) utility function; the second term is the additively separable utility function that expresses the disutility from pollution stock. An increase in pollution reduces the utility of the consumer, and increases the marginal disutility. Thus, the utility of the representative agent is influenced by the environmental level.

The accumulation equation of pollution is as follows:

$$\dot{X}(t) = x(t) - \eta X(t), \tag{2.7}$$

where η expresses the degree of purification capacity of pollution (or depollution ratio), and $\eta \in [0, 1]$.[8]

The accumulation equation of capital is

$$\dot{K}(t) = Ae^{gt}K(t)^{\alpha}L^{1-\alpha}z(t) - \delta K(t) - C(t),$$

---

[8] If $\eta = 1$, this refers to the specific case of flow-type pollution, since the pollution stock will be removed during this term.

namely,

$$k(t) = Ae^{gt}k(t)^\alpha z(t) - \delta k(t) - c(t), \tag{2.8}$$

where δ is the depreciation rate, $\delta \in [0, 1]$, and $C(t) \equiv c(t)L$ expresses the total amount of consumption.

Under the above setup, a social planner solves the following maximization problem:

$$\max \int_0^\infty e^{-\rho t}\left(\frac{c(t)^{1-\sigma} - 1}{1 - \sigma} - \frac{B}{\gamma}X(t)^\gamma\right) dt$$

$$\text{subject to} \quad \dot{k}(t) = Ae^{gt}k(t)^\alpha z(t) - \delta k(t) - c(t)$$

$$\dot{X}(t) = Ae^{gt}k(t)^\alpha Lz(t)^\beta - \eta X(t)$$

$$k(0) = k_0, \quad X(0) = X_0,$$

where ρ is a subjective discount rate and $\rho \in (0, 1)$.

The following conditions are satisfied on the optimal path to the steady state (see Appendix A.1):

$$c(t) = \lambda(t)^{-\frac{1}{\sigma}} \tag{2.9}$$

$$z(t) = \begin{cases} 1 & (\text{for } \lambda(t) \geqq \mu(t)\beta L) \\ \left(\frac{\lambda(t)}{\mu(t)\beta L}\right)^{\frac{1}{\beta-1}} & (\text{for } \lambda(t) < \mu(t)\beta L) \end{cases} \tag{2.10}$$

$$\frac{\dot{\lambda}(t)}{\lambda(t)} = \begin{cases} \rho + \delta - \alpha\left(1 - \frac{\mu(t)L}{\lambda(t)}\right)\frac{y(t)}{k(t)} & (\text{for } z(t) = 1) \\ \rho + \delta - \alpha\left(1 - \frac{1}{\beta}\right)\frac{y(t)}{k(t)} & (\text{for } z(t) < 1) \end{cases} \tag{2.11}$$

$$\frac{\dot{\mu}(t)}{\mu(t)} = \rho + \eta - \frac{BX(t)^{\gamma-1}}{\mu(t)}, \tag{2.12}$$

where $\lambda(t)$ and $\mu(t)$ represent the shadow price of capital $k(t)$ and pollution stock $X(t)$, respectively.

Next, the author examines the dynamics of each variable. As there are two state variables in this model—namely, $k(t)$ and $X(t)$—it is impossible to undertake analysis based on phase diagrams. Therefore, the author focuses on some variables of consideration. Suppose that the initial value of capital, $k(0) = k_0$, takes a smaller value than the value at the steady state. Then the shadow price $\lambda(0)$ of capital $k(0)$ is sufficiently larger than the shadow price μ of pollution stock $X(0)$—and, as a result, $z(0)$ becomes 1. $k(t)$ increases over time to the steady state. If t is sufficiently small (i.e., the economy is in an early stage of development), $k(t)$ is small and $z(t) = 1$. When $z(t) = 1$, $y(t)$, $k(t)$, and $x(t)$ also increase over time. However, as capital accumulation progresses, $\lambda(t)$ decreases and $\mu(t)$ increases. Eventually, $\mu(t)$ exceeds $\lambda(t)$, and then $z(t) < 1$. After that, $z(t)$ decreases over time. In this situation, there is the possibility of reduced pollution.

Next, the author examines the long-term growth rate of the economy, while focusing on the steady state. This economy converges the unique steady state. The growth rates of $y(t)$, $k(t)$, and $c(t)$ are the same, and given by the following formula:

$$g_y = \frac{\gamma(\beta - 1)g}{(1 - \alpha)\gamma(\beta - 1) + (\gamma + \sigma - 1)}, \tag{2.13}$$

where $g_\bullet$ expresses the growth rate with regard to the suffix variables (e.g., $g_y \equiv \dot{y}/y$). Then, the author has the following as the change rate of pollutant emissions (see Appendix A.2):

$$g_X = g_x = \frac{1 - \sigma}{\gamma} g_y. \tag{2.14}$$

Therefore, if and only if $\sigma > 1$, Eq. (2.14) derives a negative value, and pollution flow and pollution stock decrease in the long run. This implies that the EKC shows a downward-sloping curve.[9]

In summary, although pollution emissions increase with an increase in production in a country's early stage of economic development, pro-environmental production technology will be introduced and pollution emissions will decrease after economic development has reached a certain level. Moreover, both pollution flow and pollution stock decrease in the steady state, and this implies that the EKC is derived theoretically. Note that this depends on conditions regarding the inverse of the intertemporal elasticity of substitution of consumption, σ. Figure 2.4 is a pattern diagram of the EKC, derived theoretically through this model.

2.5 Problems with the Environmental Kuznets Curve

There have been numerous empirical and theoretical EKC studies. On the other hand, many criticisms have also been presented thus far—many of which argue that the EKC bears a very naive hypothesis. This section surveys the problems and criticisms of the concept, as well as academic results regarding the EKC.

2.5.1 Conceptual and Theoretical Aspects

When the EKC hypothesis is understood obediently, its main points can be summarized as follows. First, in the early stage of development, since economic growth and the environment have a trade-off relationship, some environmental pollutions are

---

[9]As is obvious from Eqs. (2.13) and (2.14), η is not included in either the long-run growth rate or the change rate of pollutant emissions. That is, whether the pollution type is flow or stock does not matter with respect to the long-run growth rate or the change rate of pollutant emissions. In addition, the depreciation rate, δ, has no impact on them, either.

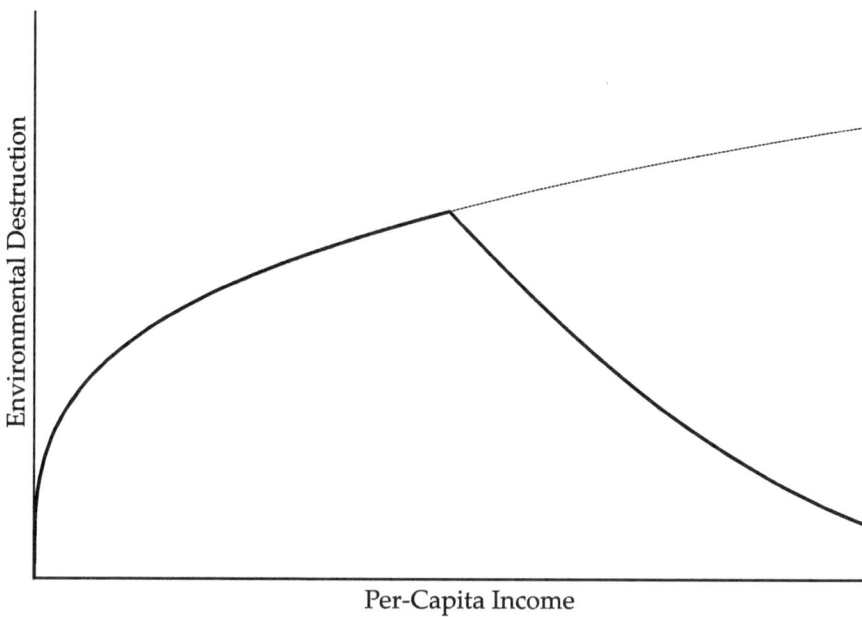

Fig. 2.4 Environmental Kuznets curve (Stokey model)

unavoidable during growth. However, that pollution may also produce permanent environmental destruction. Second, as pollution increases, the environment becomes a relatively rare resource, and so its value increases. Moreover, since appropriate measures against pollution control come to be taken with the accumulation of capital, economic growth and environmental preservation come to coexist.

One may be skeptical of these points. Arrow et al. [4], a pioneer of criticism, states that "While they [=empirical findings] do indicate that economic growth may be associated with improvements in some environmental indicators, they imply neither that economic growth is sufficient to induce environmental improvement in general, nor that the environmental effects of growth may be ignored".[10] It is quite natural that people would not rely solely on economic development and, needless to say, that certain measures by which to improve the environment would be indispensable. Moreover, other studies also point out various problems inherent in the EKC, based on the views of Arrow et al. [4]. Such criticisms and problems can be summarized as follows.

First, the interdependence of economy and the environment has not been taken into consideration. In empirical studies, income is considered an exogenous variable that affects the environment unilaterally. In reality, however, a path in the opposite direction—from the environment to the economy—can also be considered. For example, a typical case is that environmental destruction can inhibit economic growth.

---

[10]Arrow et al. [4], p. 520.

Therefore, it is necessary to consider the influence of both path directions between the economy and the environment.

Second, neither emissions of alternative pollutants nor discharges in foreign countries have been taken into consideration. Usually, reduced pollution levels are brought about by the introduction of environmental regulations and technological progress. In some situations, however, there is the possibility that reductions in a certain pollutant will bring about the emission of alternative pollutants. Moreover, tighter pollution-emission regulations in a certain country may increase emissions in a foreign country that has looser regulations. Thus, by inducing other problems, some cases that give rise to neither improvements nor sustainability may come to pass.

Third, the effects of trade have not been taken into consideration. Developing countries with relatively abundant labor and natural resources have an advantage with respect to pollution-intensive production, compared to advanced countries with abundant human resources. Suppose trade is based on a comparative advantage: developing countries with low incomes and a weak preference for the environment will specialize in pollution-intensive industries and hence increase pollution. Moreover, the tightening of regulations in an advanced nation may drive domestic pollution-intensive industries to relocate to countries with looser regulations. However, few studies analyze the influence of such leakage.

Fourth, differences in income distribution within each country have not been taken into consideration. Worldwide, income distribution is not uniform, and a large proportion of the world's population belongs to low-income developing countries. Therefore, even if the existence of the EKC were confirmed, the worldwide pollution level is expected to worsen over a considerably long period of time, if those countries located on the upward-sloping part of the EKC continue to grow economically.[11]

As to why the EKC forms, to date, not many theoretical analyses have focused on industrial structure, trade structure, or institutional aspects.[12]

2.5.2 Empirical Aspects

With respect to empirical studies of the EKC, the following have been highlighted as problematic. One large problem relates to model formulation. To date, there have been few empirical studies in line with the theoretical model. The standard regression model is a reduced-form model like Eq. (2.1), which uses per-capita income as an explanatory variable.

Hence, it is difficult to make economic interpretations of the value of the estimated parameters, and so there is a gap between the theoretical and empirical results. This type of estimation model, however, is thought to be inevitable, given the availability and quality of environmental indicator data.

---

[11] See Selden and Song [36] and Holtz-Eakin and Selden [23].

[12] Even if the EKC of CO_2 were to exist, there are many criticisms that it merely reflects energy-saving after the oil crisis, rather than emissions-reduction efforts.

Moreover, many empirical studies thus far are based on worldwide panel data; there have not been many analyses at the individual country level. Estimations with panel data assume that all countries have the same locus; however, as one can easily imagine, each country has different conditions that characterize its economic development, such as natural and social conditions. In addition, there is the possibility that for a developing country, an EKC will shift to the lower left, owing to the so-called latecomer advantage. Therefore, some researchers have posed questions regarding the methods used in panel data analysis. Dijkgraaf and Vollebergh [13] argue that the estimation of homogeneous parameters by using panel data is inappropriate, and they question the existence of the EKC.

As for modeling, the question of omitted variable bias has also been raised. Some other variables—such as trade and energy price, which may affect environmental indicators as well as income—are frequently omitted when estimating a polynomial equation of income.[13] Few studies within the EKC literature analyze changes in industrial structure or institutions. This is also considered an omitted variable bias problem, and it stems from data unavailability.

Furthermore, there is the problem of simultaneity. Although the reduced-form model implicitly assumes causality from income (on the right-hand side) to the environment (on the left-hand side), there can also be causality in the reverse direction; simultaneous determination is also possible.[14] For this reason, there is a need for an estimation method that considers simultaneity.

Another problem is heteroskedasticity. This problem generally persists in analyses of panel data, as such analyses focus on differences among individuals. As for the EKC, it is natural to consider that the social conditions of each country and which underlie the EKC will differ.

Moreover, as Stern and Common [42] point out, there is the problem of data stationarity. To date, this problem has not been overly conspicuous in the empirical EKC studies. Recently, a few research studies—such as those of Perman and Stern [33], Cole [7] (whose target is CO_2), Dinda and Coondoo [16], and Richmond and Kaufmann [34]—perform unit root tests and cointegration tests.

As mentioned, some problems persist in the estimation results reported thus far. However, even when a researcher attempts to improve the analytical methods, he or she is compelled to face the difficulties and limits inherent in empirical analysis, as the available environmental data are quite insufficient, both qualitatively and quantitatively. Therefore, based on current research trends, it is not necessarily easy to perform adequate empirical studies of environmental problems—especially of the global warming issue—while bearing in mind the issue of sustainable development.

---

[13] Suri and Chapman [44] and Cole [7, 8] each add a trade indicator to their explanatory variables. Richmond and Kaufmann [34] employ an energy consumption variable. Neumayer [30] considers natural factors such as temperature.

[14] Coondoo and Dinda [10] adopt a Granger causality test for income and CO_2 emissions, using 1960–1990 data. They report no causality, or causality from income to CO_2 emissions, in developing countries, as well as causality from CO_2 emissions to income in advanced countries.

References

1. Agras, J., & Chapman, D. (1999). A dynamic approach to the environmental Kuznets curve hypothesis. *Ecological Economics, 28*, 267–277.
2. Al-Mulali, U., Saboori, B., & Ozturk, I. (2015). Investigating the environmental Kuznets curve hypothesis in Vietnam. *Energy Policy, 76*, 123–131.
3. Andreoni, J., & Levinson, A. (2001). The simple analytics of the environmental Kuznets curve. *Journal of Public Economics, 80*, 269–286.
4. Arrow, K. J., Bolin, B., Costanza, R., Dasgupta, P., Folke, C., Holling, C. S., et al. (1995). Economic growth, carrying capacity and the environment. *Science, 268*, 520–521. Reprinted in *Ecological Economics, 15*, 91–95, in *Environment and Development Economics, 1*, 104–110, and in *Ecological Applications, 6*, 13–15.
5. Bartz, S., & Kelly, D. L. (2008). Economic growth and the environment: Theory and facts. *Resource and Energy Economics, 30*, 115–149.
6. Bulte, E. H., & van Soest, D. P. (2001). Environmental degradation in developing countries: Households and the (reverse) environmental Kuznets curve. *Journal of Development Economics, 65*, 225–235.
7. Cole, M. A. (2003). Development, trade, and the environment: How robust is the environmental Kuznets curve? *Environment and Development Economics, 8*, 557–580.
8. Cole, M. A. (2004). Trade, the pollution haven hypothesis and the environmental Kuznets curve: Examining the linkages. *Ecological Economics, 48*, 71–81.
9. Cole, M. A., Rayner, A. J., & Bates, J. M. (1997). The environmental Kuznets curve: An empirical analysis. *Environment and Development Economics, 2*, 401–416.
10. Coondoo, D., & Dinda, S. (2002). Causality between income and emission: A country group-specific econometric analysis. *Ecological Economics, 40*, 351–367.
11. Dasgupta, S., Laplante, B., Wang, H., & Wheeler, D. (2002). Confronting the environmental Kuznets curve. *Journal of Economic Perspectives, 16*, 147–168.
12. de Bruyn, S. M., & Heintz, R. J. (2002). The environmental Kuznets curve hypothesis. In J. C. J. M. van den Bergh (Ed.), *Handbook of environmental and resource economics* (pp. 656–677). Oxford: Edward Elgar.
13. Dijkgraaf, E., & Vollebergh, H. R. J. (2005). A test for parameter homogeneity in CO_2 panel EKC estimations. *Environmental and Resource Economics, 32*, 229–239.
14. Dinda, S. (2004). Environmental Kuznets curve hypothesis: A survey. *Ecological Economics, 49*, 431–455.
15. Dinda, S. (2005). A theoretical basis for the environmental Kuznets curve. *Ecological Economics, 53*, 403–413.
16. Dinda, S., & Coondoo, D. (2006). Income and emission: A panel data-based cointegration analysis. *Ecological Economics, 57*, 167–181.
17. Egli, H., & Steger, T. M. (2007). A dynamic model of the environmental Kuznets curve: Turning point and public policy. *Environmental and Resource Economics, 36*, 15–34.
18. Forster, B. A. (1973). Optimal capital accumulation in a polluted environment. *Southern Economic Journal, 39*, 544–547.
19. Galeotti, M., & Lanza, A. (1999). Richer and cleaner? A study on carbon dioxide emissions in developing countries. *Energy Policy, 27*, 565–573.
20. Galeotti, M., Lanza, A., & Pauli, F. (2006). Reassessing the environmental Kuznets curve for CO_2 emissions: A robustness exercise. *Ecological Economics, 57*, 152–163.
21. Grossman, G. M., & Krueger, A. B. (1991). Environmental impacts of a North American free trade agreement, NBER Working Paper No. 3914, NBER.
22. Grossman, G. M., & Krueger, A. B. (1995). Economic growth and the environment. *Quarterly Journal of Economics, 110*, 353–377.
23. Holtz-Eakin, D., & Selden, T. M. (1995). Stoking the fires? CO_2 emissions and economic growth. *Journal of Public Economics, 57*, 85–101.
24. John, A., & Pecchenino, R. (1994). An overlapping generations model of growth and the environment. *Economic Journal, 104*, 1393–1410.

25. Kijima, M., Nishide, K., & Oyama, A. (2010). Economic models for the environmental Kuznets curve: A survey. *Journal of Economic Dynamics and Control, 34*, 1187–1201.
26. Kuznets, S. (1955). Economic growth and income inequality. *American Economic Review, 45*, 1–28.
27. Lieb, C. M. (2002). The environmental Kuznets curve and satiation: A simple static model. *Environment and Development Economics, 7*, 429–448.
28. López, R. (1994). The environment as a factor of production: The effects of economic growth and trade liberalization. *Journal of Environmental Economics and Management, 27*, 163–184.
29. McConnell, K. E. (1997). Income and the demand for environmental quality. *Environment and Development Economics, 2*, 383–399.
30. Neumayer, E. (2002). Can natural factors explain any cross-country differences in carbon dioxide emissions? *Energy Policy, 30*, 7–12.
31. Ordás Criado, C. (2008). Temporal and spatial homogeneity in air pollutants panel EKC estimations: Two nonparametric tests applied to Spanish provinces. *Environmental and Resource Economics, 40*, 265–283.
32. Panayotou, T. (2000). Economic growth and the environment, CID Working Paper No. 56.
33. Perman, R., & Stern, D. I. (2003). Evidence from panel unit root and cointegration tests that the environmental Kuznets curve does not exist. *Australian Journal of Agricultural and Resource Economics, 47*, 325–347.
34. Richmond, A. K., & Kaufmann, R. K. (2006). Is there a turning point in the relationship between income and energy use and/or carbon emissions? *Ecological Economics, 56*, 176–189.
35. Schmalensee, R., Stoker, T. M., & Judson, R. A. (1998). World carbon dioxide emissions: 1950–2050. *Review of Economics and Statistics, 80*, 15–27.
36. Selden, T. M., & Song, D. (1994). Environmental quality and development: Is there a Kuznets curve for air pollution emissions? *Journal of Environmental Economics and Management, 27*, 147–162.
37. Selden, T. M., & Song, D. (1995). Neoclassical growth, the J curve for abatement, and the inverted U curve for pollution. *Journal of Environmental Economics and Management, 29*, 162–168.
38. Shafik, N. (1994). Economic development and environmental quality: An econometric analysis. *Oxford Economic Papers, 46*, 757–773.
39. Shafik, N., & Bandyopadhyay, S. (1992). Economic growth and environmental quality: Time series and crosscountry evidence, World Bank Policy Research Working Paper, WPS 904.
40. Stern, D. I. (1998). Progress on the environmental Kuznets curve? *Environment and Development Economics, 3*, 173–196.
41. Stern, D. I. (2004). The rise and fall of the environmental Kuznets curve. *World Development, 32*, 1419–1439.
42. Stern, D. I., & Common, M. S. (2001). Is there an environmental Kuznets curve for sulfur? *Journal of Environmental Economics and Management, 41*, 162–178.
43. Stokey, N. L. (1998). Are there limits to growth? *International Economic Review, 39*, 1–31.
44. Suri, V., & Chapman, D. (1998). Economic growth, trade and energy: Implications for the environmental Kuznets curve. *Ecological Economics, 25*, 195–208.
45. Vincent, J. R. (1997). Testing for environmental Kuznets curves within a developing country. *Environment and Development Economics, 2*, 417–431.
46. Wagner, M. (2008). The carbon Kuznets curve: A cloudy picture emitted by bad econometrics? *Resource and Energy Economics, 30*, 388–408.
47. World Bank (1992). *World development report 1992: Development and the environment.* New York: Oxford University Press.

Chapter 3
Empirical Analysis of the Environmental Kuznets Curve

Abstract This chapter undertakes an empirical analysis by using the most recent data of 171 countries, from the 1960–2010 period. The author introduces some elaborate estimation methods, in order to cope with some of the estimation problems addressed in Chap. 2—that is, the author carries out panel unit root tests and panel cointegration tests, and then estimates a dynamic panel data model, to examine the environmental Kuznets curve (EKC) hypothesis with regard to carbon dioxide. The analytical results show that: (1) the EKC is N-shaped formally but inverted U-shaped substantively, (2) the income level of the turning point of the whole world sample is approximately USD30,000 (2005 constant prices), and (3) the level of the turning point is relatively lower in advanced countries and higher in developing countries. Finally, the author briefly discusses the role of the Kyoto Protocol, based on the estimation results.

Keywords Dynamic panel data model · Panel unit roots tests · Panel cointegration tests · Turning point · Kyoto protocol

3.1 Data Sources

This section surveys the data used herein. The author employs the most recent data in a regression analysis. This is a unique feature of this study: previous empirical analyses of the environmental Kuznets curve (EKC) for carbon dioxide (CO_2) used data only to the mid-2000s. In this chapter, the author estimates the EKC for CO_2 by using the most recent available data, to 2010.

With regards to the CO_2 emissions of each country, data from the Oak Ridge National Laboratory (ORNL) were employed. These data include CO_2 emissions from fossil fuel consumption and cement production. Although CO_2 is a stock-type pollutant, the author uses not stock data but flow emission data, as Eq. (2.14) implies that there is no difference between stock and flow pollutants.[1]

---

[1] There are (1) cumulative emissions, and (2) concentration in the atmosphere to be considered as stock variables. Overall, (1) considers neither absorption nor decomposition, and (2) is not

© Development Bank of Japan 2016
K. Uchiyama, *Environmental Kuznets Curve Hypothesis and Carbon Dioxide Emissions*, Development Bank of Japan Research Series,
DOI 10.1007/978-4-431-55921-4_3

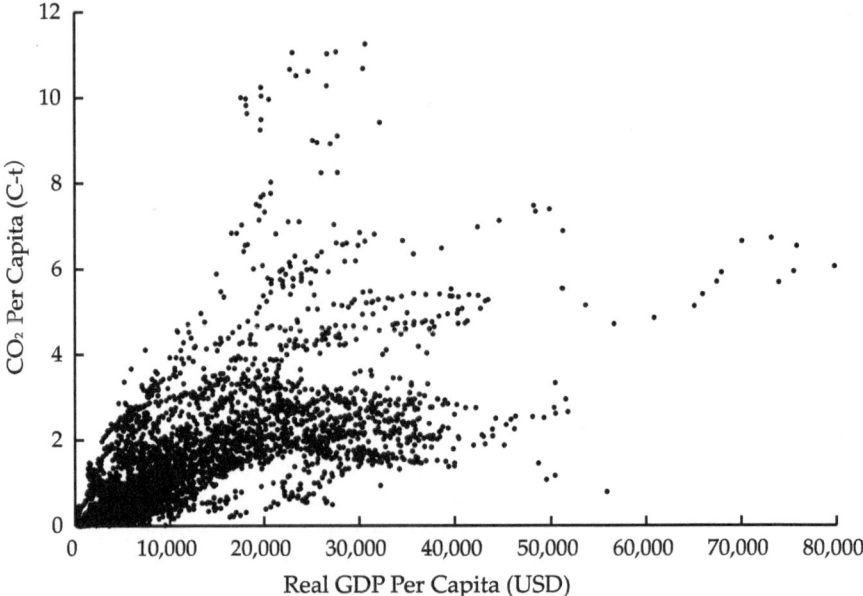

Fig. 3.1 Scatter diagram of emissions versus real GDP per capita

For per-capita income, data from the Penn World Table Version 7.1 (PWT 7.1) [8] were used. The RGDPCH series, which is measured in terms of purchasing power parity, is employed as real GDP per capita data. Moreover, the data of those countries whose ratio of CO_2 emitted in connection with oil production (i.e., gas-flaring) to total emissions is remarkably high (i.e., average of 75 % or more) were excluded from the sample.

The dataset used in this chapter is a panel dataset built using data basically from the 1960–2010 period and of 171 countries; these were obtained from ORNL and PWT 7.1. However, the author could not acquire all time-series data for some countries; therefore, the dataset is an unbalanced panel dataset. Among the 171 countries, 91 (53 %) have full and complete data from the 1960–2010 period. Based on the dataset, for the year 2010, the 171 countries represent 95 % share of the population, and cover 96 % of all CO_2 emissions worldwide. Figure 3.1 is a scatter diagram of the total observations.

Table 3.1 lists the target countries classified by the Annex of the Framework Convention on Climate Change.[2] This chapter mainly concerns itself with estimations

_____

(Footnote 1 continued)

appropriate for EKC analysis, if the researcher considers it a constant everywhere worldwide. For these reasons, the analyses in this chapter use flow data series, as has been done in previous studies.

[2]Most previous studies classify countries in terms of OECD membership. This classification is appropriate, since the EKC is used to investigate the relationships in economic development. The author, however, adopts the classification of the Framework Convention on Climate Change, as the author is conscious of the political backgrounds of international negotiations on the global warming

Table 3.1 Countries classified as per the Framework Convention on Climate Change

Annex I (40 countries)
Annex II (23 countries)
Australia*, Austria*, Belgium*, Canada*, Denmark*, Finland*, France*, Germany*, Greece*, Iceland*, Ireland*, Italy,*, Japan*, Luxembourg*, Netherlands*, New Zealand*, Norway*, Portugal*, Spain*, Sweden*, Switzerland*, United Kingdom*, United States of America*
Economy in Transition (14 countries)
Belarus, Bulgaria, Croatia, Czech Republic*, Estonia*, Hungary*, Latvia, Lithuania, Poland*, Romania, Russian Federation, Slovakia*, Slovenia*, Ukraine
Others (3 countries)
Cyprus, Malta, Turkey
Non-Annex I (131 countries)
Afghanistan, Albania, Algeria, Angola, Antigua and Barbuda, Argentina, Armenia, Azerbaijan, Bahamas, Bahrain, Bangladesh, Barbados, Belize, Benin, Bhutan, Bolivia, Bosnia and Herzegovina, Botswana, Brazil, Burkina Faso, Burundi, Cambodia, Cameroon, Cape Verde, Central African Republic, Chad, Chile*, China, Colombia, Comoros, Congo, Costa Rica, Cote d'Ivoire, Cuba, Djibouti, Dominica, Dominican Republic, Ecuador, Egypt, El Salvador, Equatorial Guinea, Eritrea, Ethiopia, Fiji, Gabon, Gambia, Georgia, Ghana, Grenada, Guatemala, Guinea, Guinea-Bissau, Guyana, Haiti, Honduras, India, Indonesia, Iran, Iraq, Israel*, Jamaica, Jordan, Kazakhstan, Kenya, Kiribati, Kyrgyzstan, Laos, Lebanon, Liberia, Macedonia, Madagascar, Malawi, Malaysia, Maldives, Mali, Marshall Islands, Mauritania, Mauritius, Mexico*, Federated States of Micronesia, Moldova, Mongolia, Morocco, Mozambique, Namibia, Nepal, Nicaragua, Niger, Pakistan, Palau, Panama, Papua New Guinea, Paraguay, Peru, Philippines, Republic of Korea*, Rwanda, Saint Kitts and Nevis, Saint Lucia, Saint Vincent and the Grenadines, Samoa, Sao Tome and Principe, Senegal, Seychelles, Sierra Leone, Singapore, Solomon Islands, Somalia, South Africa, Sri Lanka, Sudan, Suriname, Swaziland, Syria, Tajikistan, Tanzania, Thailand, Togo, Tonga, Trinidad and Tobago, Tunisia, Turkmenistan, Uganda, Uruguay, Uzbekistan, Vanuatu, Venezuela, Vietnam, Yemen, Zambia, Zimbabwe

Note "∗" expresses OECD members

of the world EKC by using the whole sample; it also contains some sub-sample analyses.

The descriptive statistics are shown in Table 3.2. In looking at the averages, the Annex II countries—which comprise the group of advanced countries—have high CO_2 emissions and real GDP values. However, the CO_2 distribution seems to be biased, as there is deviation between the average and median values. Although the non-Annex I countries (131 countries) comprise the group of developing countries, they also include comparatively high-income countries such as Israel and Singapore. Therefore, there is great bias in the distribution of both CO_2 and real GDP, with large deviations between the maximum and median values. Moreover, since most of the countries that are transitioning to a market economy (14 countries) have nothing more than data from 1992 and later, the analysis here may contain a certain degree of bias.

(Footnote 2 continued)

issue; these negotiations are held by the groups based on the Framework. This method does not differ much from those of previous studies, as most Annex I countries have joined the OECD.

Table 3.2 Descriptive statistics

	Average	Median	Maximum	Minimum	Standard deviation	Coefficient of variation	Observations
CO_2 per capita (CO2PC): carbon-t/person							
Total (171 countries)	0.93	0.35	11.25	0.00	1.31	1.41	7,319
Annex I (40 countries)	2.40	2.09	11.07	0.16	1.46	0.61	1,644
Annex II (23 countries)	2.67	2.27	11.07	0.25	1.59	0.59	1,142
Economy in Transition (14 countries)	2.06	1.97	4.11	0.72	0.76	0.37	359
Non-Annex I (131 countries)	0.50	0.20	11.25	0.00	0.89	1.76	5,675
Real GDP per capita (RGDPPC): USD/person							
Total (171 countries)	7,884	3,899	80,231	161	9,622	1.22	7,319
Annex I (40 countries)	19,962	18,507	80,231	1,371	11,126	0.56	1,644
Annex II (23 countries)	24,256	22,938	80,231	4,154	10,321	0.43	1,142
Economy in Transition (14 countries)	10,264	9,281	27,141	1,371	4,909	0.48	359
Non-Annex I (131 countries)	4,385	2,520	55,862	161	5,390	1.23	5,675

Note Real GDP per capita is based on 2005 constant prices

3.2 Stationarity of Panel Data

3.2.1 Panel Unit Roots Tests

Many empirical EKC analyses use panel data. Whenever panel data reflect a considerably long period of time-series data, it is necessary to examine stationarity. However, few studies to date have paid attention to data stationarity. When panel data are constructed with nonstationary variables, it is possible during analysis that uses the panel that spurious regressions will occur, as is the case with ordinary time-series analysis. However, standard unit root tests for each individual are problematic, in that they are apt to accept the null hypothesis that the data is nonstationary, due to limited power.[3] As methods for unit root tests with panel data have been developed, some studies have introduced them to examine data stationarity, in advance of their

---

[3] Standard individual unit root tests have less explanatory power than panel unit root tests, mainly because of the small number of time-series observations.

analyses. Perman and Stern [16], in an analysis of sulfur excretions, employed a panel unit root test for each variable used in the regression analysis, and obtained the result that the null hypothesis that the series contains a unit root cannot be rejected. Cole [4] applied panel unit root tests to CO_2 emissions, income, and other explanatory variables, and found that the null hypothesis cannot be rejected, and also that the first-order difference series is stationary (i.e., I(1)). Romero-Ávila [18] and Wagner [21] also carried out panel unit roots tests for CO_2.

The author undertook panel unit root tests for the variables of the sample 171 countries used in the analysis. Some methods are proposed as panel unit root tests: the Levin et al. [12] (LLC) test, which is an extension of that of Levin and Lin [11]; the Im, Pesaran and Shin [10] (IPS) test; the Fisher-type augmented Dickey–Fuller (F-ADF) test of Maddala and Wu [13] and Choi [3]; and the test of Hadri [6].

In this section, the author undertakes three types of test (i.e., LLC, IPS, and F-ADF), along with robustness checks.[4] Panel unit root tests are roughly classified into two types, based on differences of assumption: LLC is one type, and IPS and F-ADF are another (see Appendix B.1). The author employs the model with a constant term (individual effects) and trend for the tests. The results are shown in Table 3.3. Note that in that table, CO2PC denotes CO_2 per capita, and RGDPPC is real GDP per capita.

First, in looking at the test results concerning the level data of CO2PC (upper part of Table 3.3), one finds that all the test results conclude that the null hypothesis (i.e., that each series in the panel contains a unit root) cannot be rejected—not even at the 10 % significance level. By taking the first-order difference, the author can reject the null hypothesis and find each series of the panel is a stationary process (lower part of Table 3.3). Second, in looking at the test results concerning the level data of RGDPPC, $(RGDPPC)^2$, and $(RGDPPC)^3$, none of the tests can reject the null hypothesis; this suggests the existence of a unit root. In testing the first-order difference series, however, the null hypothesis can be rejected, and this implies that each series in the panel is a stationary process. Based on the above test results, in the following analyses, the author assumes each variable is I(1).

3.2.2 Panel Cointegration Tests

Since each variable used in this section is nonstationary, the regression analysis using the level data becomes spurious. Therefore, based on the assumption that each variable is I(1), the author investigates whether or not the relation in Eq. (2.1) comes into existence as the long-term stable equilibria, by applying panel cointegration tests. When a long-term equilibrium relation can be confirmed, the regression is not

---

[4]Perman and Stern [16] and Richmond and Kaufmann [17] each employ LLC and IPS, and Cole [4] and Dinda and Coondoo [5] each apply IPS.

Table 3.3 Panel unit root tests

		Individual linear trends		Individual linear trends	
		Yes		No	
		Statistics	p-value	Statistics	p-value
Level data					
CO2PC	LLC	0.589	0.722	1.151	0.875
	IPS	1.035	0.850	2.247	0.988
	F-ADF	339.640	0.526	370.530	0.139
RGDPPC	LLC	6.086	1.000	8.828	1.000
	IPS	5.152	1.000	15.540	1.000
	F-ADF	293.125	0.974	176.869	1.000
$(RGDPPC)^2$	LLC	12.401	1.000	18.448	1.000
	IPS	11.567	1.000	22.142	1.000
	F-ADF	245.469	1.000	144.970	1.000
$(RGDPPC)^3$	LLC	17.619	1.000	24.418	1.000
	IPS	16.765	1.000	25.665	1.000
	F-ADF	227.952	1.000	149.710	1.000
First-order difference data					
CO2PC	LLC	−27.263	0.000	−31.644	0.000
	IPS	−40.120	0.000	−44.461	0.000
	F-ADF	2220.990	0.000	2610.810	0.000
RGDPPC	LLC	−28.342	0.000	−26.701	0.000
	IPS	−31.153	0.000	−34.607	0.000
	F-ADF	1671.080	0.000	1967.080	0.000
$(RGDPPC)^2$	LLC	−23.762	0.000	−20.632	0.000
	IPS	−29.773	0.000	−30.927	0.000
	F-ADF	1638.490	0.000	1831.070	0.000
$(RGDPPC)^3$	LLC	−18.106	0.000	−14.314	0.000
	IPS	−28.554	0.000	−28.010	0.000
	F-ADF	1650.300	0.000	1761.470	0.000

Note Null hypothesis: Variable contains a unit root. LLC: Levin et al. [12] test. IPS: Im, Pesaran and Shin [16] test. F-ADF: Fisher-type ADF (Maddala and Wu [13], Choi [3]) test

considered spurious, even if each variable is nonstationary. Some methods have been proposed with regards to panel cointegration tests (see Appendix B.2). As is the case with panel unit root tests, panel cointegration tests are more powerful than individual cointegration tests.[5]

---

[5]It has been pointed out that individual cointegration tests with a small sample have difficulties in detecting long-term and stable relations, due to the large deviation of statistics.

Table 3.4 Panel cointegration tests

Variables: CO2PC, RGDPPC, $(RGDPPC)^2$, $(RGDPPC)^3$

		Statistics
1	Panel v	9.2161 (0.000)
2	Panel ρ	−6.4627 (0.000)
3	Panel t (nonparametric)	−16.1100 (0.000)
4	Panel t (parametric)	−8.7066 (0.000)
5	Group ρ	0.2034 (0.5806)
6	Group PP (nonparametric)	−16.7050 (0.000)
7	Group ADF (parametric)	−4.6204 (0.000)

Note Null hypothesis: No cointegration

Both individual intercept and individual trebd are included for the tests

Figures in parentheses are p-values

The lag length for the ADF test is 1

See Appendix B.2 for an outline of the tests

In the analyses, the author employs the test of Pedroni [15], which is frequently used in panel cointegration tests.[6] The lag length in the ADF test is set to 1. Table 3.4 presents the test results.

The results show that the null hypothesis of no cointegration is rejected at the 5 % significant level, except for the group ρ statistics. This finding supports the assertion that there is a cointegrating relationship among CO2PC, RGDPPC, $(RGDPPC)^2$, and $(RGDPPC)^3$, and that these variables are on the long-term stable equilibria.

Based on the above results, the author considers the EKC a long-term relationship among the variables, and uses the level data to estimate the EKC.

3.3 Estimation of the Environmental Kuznets Curve

3.3.1 Estimation Methods

As mentioned, in the literature, there have been few structural estimation studies that make use of theoretical models. This section does not touch upon structural estimation, but rather estimates the EKC based on the reduced form of Eq. (2.1).

---

[6]Richmond and Kaufmann [17] also adopt the Pedroni [15] test.

By estimating the polynomial model of income, there may be the advantage that analytical results could be comparable to previous ones.

In the following, the author estimates the EKC using two models—namely, a static panel data model, and the dynamic panel data model of Arellano and Bond [2].

3.3.2   Estimation (1): Static Panel Data Model

In early studies, static panel data models were often used. This simple model may be somewhat problematic when applied to EKC estimations. However, the author employs this standard model with the aforementioned dataset, to facilitate comparisons with the results of previous studies.

The estimation model used here is a one-way error component model with cubic term y_{it}^3:

$$E_{it} = \beta_0 + \beta_1 y_{it} + \beta_2 y_{it}^2 + \beta_3 y_{it}^3 + \mu_i + u_{it}, \tag{3.1}$$

where E denotes CO2PC, y expresses RGDPPC, μ_i is individual effects, and u_{it} is the error term. Moreover, i denotes countries, and t is time ($i = 1, 2, \ldots, N$; $t = 1, 2, \ldots, T_i$). T_i is the amount of observable time for country i. Unobservable individual effects are introduced, to control for certain factors (e.g., high-income countries tend to be located in high latitudes).

The author undertook the estimation of this model by using level data. Only RGDPPC is employed here as an explanation variable; this is based on the suggestion by Stern [19], Agras and Chapman [1], and Neumayer [14] that per-capita income has the greatest explanatory power. Another reason is that many of the early studies used only RGDPPC as an explanatory variable. The expected sign conditions are $\beta_1 > 0$, $\beta_2 < 0$, and $\beta_3 = 0$.

In the following, the author mainly focuses on the estimation results from using the entire sample of 171 countries (hereafter, "the whole world"). Furthermore, the author also undertook estimations with subsamples such as Annex I, Annex II, and non-Annex I countries, in order to check the robustness of the estimations.

The estimation results of the fixed effects model are shown in Table 3.5, which omits the report of individual effects. The estimated parameter of (RGDPPC)^3 is positive and statistically significant for the whole world and each of the Annex I and Annex II groups. This implies that the expected sign conditions in Eq. (2.1) are not satisfied and that the estimated EKC is an N-shaped curve. In addition, the parameter of (RGDPPC)^3 of non-Annex I countries is estimated as being negative, with statistical significance. This means the EKC is an inverted N-shaped curve.

Table 3.5 Estimation results (1): Static panel data (fixed effects) model

Group	Whole world	Annex I	Annex II	Non-Annex I
Constant	0.054496***	0.752608***	0.061686	0.070868***
	(0.017952)	(0.082799)	(0.130683)	(0.015043)
RGDPPC	0.000174***	0.000163***	0.000242***	0.0000749***
	(4.15E−06)	(9.56E−06)	(1.35E−05)	(5.69E−06)
$(RGDPPC)^2$	−3.67E−09***	−3.70E−09***	−5.95E−09***	4.37E−09***
	(1.48E−10)	(3.05E−10)	(4.12E−10)	(3.37E−10)
$(RGDPPC)^3$	1.67E−14***	1.89E−14***	3.68E−14***	−1.13E−13***
	(1.49E−15)	(2.78E−15)	(3.59E−15)	(5.22E−15)
$Adj.R^2$	0.931	0.900	0.905	0.898
Turning Point	29,745	28,059	27,386	32,566
Countries	171	40	23	131
Observations	7,319	1,644	1,142	5,675

Note Figures in parentheses are standard errors
***, ** and * indicate significance at the 1, 5 and 10 % levels, respectively
The unit of the turning point is USD (2005 constant prices)

   In the following, the author interprets the concave downward point of the two inflection points of a cubic function as the turning point.[7] The turning point of the whole world is USD29,745 (2005 constant prices), and this is within the range of the sample.

   The turning point of the Annex II group is the minimum of USD27,386, and that of Annex I is USD28,059; indeed, the difference is very small. Each group has a turning point within the sample, and the level is found to be close to those seen in previous studies. As mentioned, the Annex II group consists of Annex I countries and those countries that are transitioning to a market economy. It is thought that the countries in transition take the upper shift of the turning point; however, it is important to pay attention to evaluations of these results, given the small country samples.

   Although one needs to bear in mind that non-Annex I countries, or developing countries, tend to have an inverted N-shaped curve, their turning point is calculated as USD32,566—the highest among these groups. However, in comparing these results to those of some previous studies that concluded substantively about the monotonic increase of EKC among developing countries, one finds that the numerical values of this work are smaller and within a reasonable range.

---

[7]As for Eq. (3.1), the turning point is calculated according to

$$\frac{-\beta_2 - \sqrt{\beta_2^2 - 3\beta_3\beta_1}}{3\beta_3} \quad (\text{for } \beta_3 > 0), \quad \text{and} \quad \frac{-\beta_2 + \sqrt{\beta_2^2 - 3\beta_3\beta_1}}{3\beta_3} \quad (\text{for } \beta_3 < 0).$$

Note that this static panel data model has some technical problems, as pointed out in Sect. 2.5. There is the possibility that the estimation result of the inverted N-shaped EKC for the non-Annex I group relates to problems arising from the estimation models and methods themselves.

3.3.3 Estimation (2): Dynamic Panel Data (Arellano and Bond) Model

The relationship between CO_2 emissions and income is considered a long-term stable equilibrium (see Sect. 3.2.2); therefore, if the EKC is approved, it implies an adjustment process to a long-term equilibrium—a process that is thought to require a certain amount of time. In this case, the model with time factors is applicable to the estimation [16]. Static models like Eq. (3.1), on the other hand, are considered appropriate when the adjustment speed is very fast.

Moreover, as stated in Sect. 2.5, the model in Eq. (3.1) is very problematic, in that it does not take into account simultaneity. In a reduced form—as seen in many previous studies—explanatory variables are restricted to those that satisfy exogeneity, such as predetermined endogenous variables and exogenous variables. However, the exogeneity condition is not satisfied under simultaneity.

A generalized method of moments (GMM) estimation with a dynamic model and instrumental variables, as one appropriate model, can be applied to these problems. If researchers can clarify the dynamic process of the EKC, it may become possible to generate accurate predictions. In previous studies, Agras and Chapman [1] took the dynamic process into consideration, and applied the autoregressive distributed lag (ARDL) model. Halkos [7] applied the dynamic panel data model of Arellano and Bond [2] to estimate the EKC for sulfur emissions. Huang et al. [9] also applied the Arellano-Bond dynamic panel data model, to analyze the relationship between energy use and GDP.

The method of Arellano and Bond [2] offers one solution to a problem that arises from the static panel data estimation of the EKC. On the other hand, only a few studies in the literature touch upon EKC estimations for CO_2. In this section, the author tries to adopt this dynamic panel data model. Estimation by GMM makes it unnecessary to distinguish fixed effects and random effects, or to specify the distribution of the error term. Hence, by employing this model, researchers can enjoy some advantage in terms of high estimation stringency.

The estimation model is as follows:

$$E_{it} = \alpha E_{i,t-1} + \beta_1 y_{it} + \beta_2 y_{it}^2 + \beta_3 y_{it}^3 + \mu_i + u_{it}. \tag{3.2}$$

One feature of dynamic relationships is the presence of a lagged dependent variable among the explanatory variables. This model is estimated by GMM and thus requires applicable instrumental variables.

The estimation results are reported in Table 3.6. The results show that all estimated parameters are statistically significant at the 1 % level, and that $\beta_3 > 0$ for all groups, including the non-Annex I group. This implies that the EKC takes an N-shaped curve.

The Arellano-Bond serial correlation test and Sargan test were performed for the model specification (see Appendix C).

The Arellano-Bond serial correlation test determines that u_{it} does not have auto-correlation of a second order or more. This is a very important test. If u_{it} has autocor-relation of a second order or more, it means there are variables that cannot be used as instrumental variables—in which case, dynamic panel data estimation would be rendered inappropriate. The null hypothesis is that there is no autocorrelation. For all groups, the p-value of the AR(1) test statistics indicate significance at less than 10 %, and those of the AR(2) test at more than 10 %. These results indicate that there is no second-order autocorrelation, and hence no evidence of model misspecification.

The Sargan test is used to test for over-identifying restrictions with regard to instrumental variables; the null hypothesis is that the over-identifying restrictions are valid.[8] The test statistics indicate that for all groups, the null hypothesis is not rejected. This implies that the over-identifying restrictions are valid with regard to the instrumental variables.

The turning point of the whole world is USD30,397 (2005 constant prices).[9] Those of other groups are within the range of USD28,374 (Annex II) to USD33,751 (non-Annex I). The turning-point level is almost identical to that of the static panel data model (Table 3.5) and of many previous studies.[10]

Since the N-shaped EKC is estimated for all groups, the author calculates the income level of the concave upward point of two inflection points, for reference. When income exceeds this level, CO_2 emissions will start to increase, at which point there will be difficulties in building a sustainable society. Achieving an understand-ing of the N-shaped EKC—that is, the interpretation of the situation where CO_2 emissions increase along with an increase in income—is difficult. If the income level of the concave upward point is very high, however, researchers may be able to understand that the EKC is a substantively inverted U-shape. The calculated level is USD79,741 for the whole world, USD67,562 for Annex I, USD60,244 for Annex II, and USD432,440 for non-Annex I. If it were possible for each country to reach the income level related to its respective group, it is thought that it would occur in the far and distant future.

Although the basic estimation model of the EKC is a polynomial equation of income, some researchers assert that variables that relate to environmental indicators should be added (see Sect. 2.5). Therefore, in estimating the EKC, the author employs three additional regressors as control variables. First, the share of service sector in

---

[8]Arellano and Bond [2] point out that the Sargan test has the tendency to reject the null hypothesis excessively when disturbance has heteroskedasticity.

[9]The formula used for this calculation is the same as that in Sect. 3.3.2.

[10]Halkos [7]—who researched sulfur dioxide emissions by using Arellano and Bond's [2] model—obtained results featuring a lower level of turning point than that seen in previous research (i.e., Stern and Common [20]).

GDP (SERVICES) is employed as a variable to express industrial structure change. If the industrial structure shifts from manufacturing to services, CO_2 emissions may decrease. For this reason, the expected sign is negative. Second, the share of fossil fuels in total fuel consumption (FOSSIL_FUEL) is employed; this illustrates that differences of fuel consumption structure may influence CO_2 emissions, and its expected sign is positive. Third, the author introduces a variable to represent the effects of trade—namely, the share of trade volume (i.e., 'exports + imports') in GDP (TRADE). It is possible for a country with a high degree of dependence on foreign trade to emit smaller amounts of CO_2 domestically, as other foreign countries "cover" for the decline in CO_2 emissions. The expected sign of this variable is negative.

These variables are calculated from the World Bank's World Development Indicators. Given the state of data availability in creating these three variables, the sample size needs to be smaller than the whole world (i.e., the new sample has 118 countries).[11]

The estimation results are in the rightmost column of Table 3.6. The results show that the parameters of the additional three control variables are statistically significant, and all signs are as expected. Both the Arellano-Bond and Sargan tests generate good results with respect to the model specification. Looking at the estimated parameters, one can see that the value of FOSSIL_FUEL is larger than that of the others. This suggests that fossil fuel consumption has a relatively larger effect on the formation of the EKC, while the effect of trade is relatively small. However, the sign of the estimated β_3 is positive and statistically significant—that is, the EKC is the same N-shaped curve. The value of the estimated parameters RGDPPC, $(RGDPPC)^2$, and $(RGDPPC)^3$, as well as the income level of the turning point, are all similar to the other results in the table.[12] Therefore, based on these estimation results, the author finds that income has strong explanatory power with regard to the formation of the EKC, even though the effects of the control variables are significant.

---

[11] The sample contains the following 118 countries: Australia, Austria, Belgium, Canada, Denmark, Finland, France, Germany, Greece, Iceland, Ireland, Italy, Japan, Luxembourg, Netherlands, New Zealand, Norway, Portugal, Spain, Sweden, Switzerland, United Kingdom, United States of America, Belarus, Bulgaria, Croatia, Czech Republic, Estonia, Hungary, Latvia, Lithuania, Poland, Romania, Russian Federation, Slovakia, Slovenia, Ukraine, Cyprus, Malta, Turkey, Albania, Algeria, Angola, Argentina, Armenia, Azerbaijan, Bahrain, Bangladesh, Benin, Bolivia, Bosnia and Herzegovina, Botswana, Brazil, Cambodia, Cameroon, Chile, China, Colombia, Congo, Costa Rica, Cote d'Ivoire, Cuba, Dominican Republic, Ecuador, Egypt, El Salvador, Eritrea, Ethiopia, Gabon, Georgia, Ghana, Guatemala, Honduras, India, Indonesia, Iran, Jamaica, Jordan, Kazakhstan, Kenya, Kyrgyzstan, Lebanon, Macedonia, Malaysia, Mexico, Moldova, Mongolia, Morocco, Mozambique, Namibia, Nepal, Nicaragua, Pakistan, Panama, Paraguay, Peru, Philippines, Republic of Korea, Senegal, Singapore, South Africa, Sri Lanka, Sudan, Syria, Tajikistan, Tanzania, Thailand, Togo, Trinidad and Tobago, Tunisia, Turkmenistan, Uruguay, Uzbekistan, Venezuela, Vietnam, Yemen, Zambia, Zimbabwe.

[12] The calculated income level of the concave upward point of the inflection points is USD91,188.

Table 3.6 Estimation results (2): Dynamic panel data (Arellano and Bond) model

Group	Whole world	Annex I	Annex II	Non-Annex I	Whole world
$CO2PC_{-1}$	0.636695***	0.780403***	0.787119***	0.504199***	0.809958***
	(5.03E−07)	(0.008633)	(0.002543)	(0.001016)	(2.34E−04)
RGDPPC	5.81E−05***	5.30E−05***	8.41E−05***	6.13E−05***	5.34E−05***
	(3.88E−09)	(4.380E−06)	(2.52E−05)	(1.52E−06)	(1.41E−06)
$(RGDPPC)^2$	−1.32E−09***	−1.28E−09***	−2.18E−09***	−9.79E−10***	−1.16E−09***
	(1.13E−13)	(8.94E−11)	(7.37E−10)	(6.28E−11)	(3.52E−11)
$(RGDPPC)^3$	7.99E−15***	8.76E−15***	1.64E−14***	1.40E−15***	6.34E−15***
	(1.02E−18)	(5.79E−16)	(5.96E−15)	(6.95E−16)	(2.56E−16)
SERVICES					−0.004416***
					(6.27E−06)
FOSSIL_FUEL					0.0162***
					(2.01E−05)
TRADE					−0.000699***
					(4.46E−06)
Arellano-Bond serial correlation test					
AR(1)	−3.002478	−3.409353	−2.904396	−2.255246	−3.14695
(p-value)	(0.0027)	(0.0007)	(0.0037)	(0.0241)	(0.0016)
AR(2)	−0.986657	0.308945	0.627572	−1.212352	−0.838588
(p-value)	(0.3238)	(0.7574)	(0.5303)	(0.2254)	(0.4017)
Sargan test	148.7341	30.67648	20.95528	85.74087	81.61411
(p-value)	(0.3542)	(0.3316)	(0.3393)	(0.1654)	(0.28129)
Turning point	30.397	29.850	28,374	33,751	30,789
Countries	171	40	23	131	118
Observations	6,977	1,564	1,096	5,413	3,137

Note Figures in parentheses are standard errors

***, ** and * indicate significance at the 1, 5 and 10 % levels, respectively

The Arellano-Bond test is the test for zero autocorrelation in first-differenced errors (null: no autocorrelation)

The Sargan test is the test for overidentifying restrictions (null: the overidentifying restrictions are valid)

The unit of the turning point is USD (2005 constant prices)

3.4 Some Comments on the Turning Point

This section focuses on the relationship between the estimated turning points and the Kyoto Protocol, based on an understanding that the turning point could possibly reflect some sort of behavioral changes in the world after 1997, when the Kyoto Protocol was adopted. In other words, the author's interest is in whether or not the adoption and ratification of the Kyoto Protocol have functioned as an institution to reduce CO_2 emissions or reduce the turning-point level. The introduction of institutions to internalize external diseconomies is thought to have been a factor that can create or shift the turning point. In the following, the author briefly looks at the

Table 3.7 Changes in turning points (Whole world)

(USD)

Estimation period	Average RGDPPC	Turning point
1960–1997	6,791	35,251
1960–1998	6,862	35,037
1960–1999	8,952	33,951
1960–2000	8,809	32,638
1960–2001	8,653	28,099
1960–2002	8,585	27,793
1960–2003	8,440	27,949
1960–2004	8,516	26,678
1960–2005	8,359	27,048
1960–2006	8,248	28,345
1960–2007	8,149	32,165
1960–2008	8,095	30,509
1960–2009	8,030	29,507
1960–2010	7,884	30,397

Note 2005 constant prices
The turning points values are calculated by using the full sample of 171 countries

issue of structural change in the EKC, by using estimation results based on data until 2010—data that have longer periods than those seen in previous studies.

The Chow test should be employed essentially to examine structural change. However, the author needed to abandon it, as it is difficult to ensure a sufficient degree of freedom in the sample after 1997—the year in which the sample is divided into two distinct samples. Therefore, the author employed an alternative method: repeating dynamic panel estimations by changing by one year the end of the estimation period between 1997 and 2010, using the sample of 171 countries. The author then examined changes in the turning points thus obtained. The changes in the turning points are shown in Table 3.7.

The Kyoto Protocol came into effect in February 2002; the United States—the world's largest greenhouse gas emitter at the time—withdrew from negotiations in 2001. Meanwhile, international society was growing suspicious about the effects of the Kyoto Protocol. In line with this, the turning point declined until 2004 and then later turned to increase. This implies that, worldwide, actions with regards to reducing CO_2 emissions worsened rather than improved following the enforcement of the Kyoto Protocol; it is possible that the effects of the Kyoto Protocol as an institution are very much limited, from the viewpoint of changes to the turning point of the EKC.

However, this may be nothing more than conjecture. This analysis cannot draw any definitive conclusion on the matter, and it remains a topic to be resolved in future research.

References

1. Agras, J., & Chapman, D. (1999). A dynamic approach to the environmental Kuznets curve hypothesis. *Ecological Economics, 28*, 267–277.
2. Arellano, M., & Bond, S. (1991). Some tests of specification for panel data: Monte Carlo evidence and an application to employment equations. *Review of Economic Studies, 58*, 277–297.
3. Choi, I. (2001). Unit root tests for panel data. *Journal of International Money and Finance, 20*, 249–272.
4. Cole, M. A. (2003). Development, trade, and the environment: How robust is the environmental Kuznets curve? *Environment and Development Economics, 8*, 557–580.
5. Dinda, S., & Coondoo, D. (2006). Income and emission: A panel data-based cointegration analysis. *Ecological Economics, 57*, 167–181.
6. Hadri, K. (2000). Testing for stationarity in heterogeneous panel data. *Econometric Journal, 3*, 148–161.
7. Halkos, G. E. (2003). Environmental Kuznets curve for sulfur: Evidence using GMM estimation and random coefficient panel data models. *Environment and Development Economics, 8*, 581–601.
8. Heston, A., Summers, R. & Aten, B. (2012). Penn world table version 7.1. Center for International Comparisons of Production, Income and Prices at the University of Pennsylvania.
9. Huang, B.-N., Hwang, M. J., & Yang, C. W. (2008). Causal relationship between energy consumption and GDP growth revisited: A dynamic panel data approach. *Ecological Economics, 67*, 41–54.
10. Im, K. S., Pesaran, M. H., & Shin, Y. (2003). Testing for unit roots in heterogeneous panels. *Journal of Econometrics, 115*, 53–74.
11. Levin, A., & Lin, C.-F. (1993). Unit root tests in panel data: New results. Discussion Paper No. 93-56, Department of Economics, University of California at San Diego.
12. Levin, A., Lin, C.-F., & Chu, C. S. J. (2002). Unit root tests in panel data: Asymptotic and finite-sample properties. *Journal of Econometrics, 108*, 1–24.
13. Maddala, G. S., & Wu, S. (1999). A comparative study of unit root tests with panel data and a new simple test. *Oxford Bulletin of Economics and Statistics, 61*, 631–652.
14. Neumayer, E. (2002). Can natural factors explain any cross-country differences in carbon dioxide emissions? *Energy Policy, 30*, 7–12.
15. Pedroni, P. (1999). Critical values for cointegration tests in heterogeneous panels with multiple regressors. *Oxford Bulletin of Economics and Statistics, 61*, 653–678.
16. Perman, R., & Stern, D. I. (2003). Evidence from panel unit root and cointegration tests that the environmental Kuznets curve does not exist. *Australian Journal of Agricultural and Resource Economics, 47*, 325–347.
17. Richmond, A. K., & Kaufmann, R. K. (2006). Is there a turning point in the relationship between income and energy use and/or carbon emissions? *Ecological Economics, 56*, 176–189.
18. Romero-Ávila, D. (2008). Questioning the empirical basis of the environmental Kuznets curve for CO_2: New evidence from a panel stationarity test robust to multiple breaks and cross-dependence. *Ecological Economics, 64*, 559–574.
19. Stern, D. I. (1998). Progress on the environmental Kuznets curve? *Environment and Development Economics, 3*, 173–196.
20. Stern, D. I., & Common, M. S. (2001). Is there an environmental Kuznets curve for sulfur? *Journal of Environmental Economics and Management, 41*, 162–178.
21. Wagner, M. (2008). The carbon Kuznets curve: A cloudy picture emitted by bad econometrics? *Resource and Energy Economics, 30*, 388–408.

Chapter 4
Concluding Remarks

Abstract This short chapter summarizes the features, obtained results, and remaining subjects of research. To achieve sustainable development and a sustainable society, it is necessary to shift the turning point of the environmental Kuznets curve (EKC) to the lower left. With this in mind, this chapter points out the necessity of offering development assistance and transferring technology to developing countries. Finally, the author touches briefly on the stationary state society, as presented by John Stuart Mill, as an image of a sustainable society.

Keywords Development assistance · Environmental technology development · John Stuart Mill · Stationary state · Sustainability

4.1 Summary of the Study

The features and results of the Brief can summarized as follows.

The first of this study's features to differentiate it from previous studies is that it employs a dynamic panel data model in its estimation method; additionally, it addresses some technical problems that relate to estimation, such as simultaneity bias. Second, the author performs panel unit root tests and panel cointegration tests, to address the stationarity of the panel data; these tests have seldom been used in previous studies. Third, the author utilizes the most recent data (i.e., up to 2010) and examines changes in the environmental Kuznets curve (EKC) or the effects of the Kyoto Protocol, from the viewpoint of turning points in income.

A number of results were derived. First, the author confirmed an N-shaped EKC. Second, the author calculated the income level at the turning point as being approximately USD30,000 (2005 constant prices) for the whole world; this figure resembles that seen in many previous studies.

Moreover, based on the estimation results, the author found that, for the whole world, the income level at the turning point decreased following the adoption of the Kyoto Protocol, but that it has turned and increased in recent years. The author inferred that the introduction of the Kyoto Protocol has had a limited impact on global warming countermeasures worldwide.

© Development Bank of Japan 2016
K. Uchiyama, *Environmental Kuznets Curve Hypothesis and Carbon Dioxide Emissions*, Development Bank of Japan Research Series, DOI 10.1007/978-4-431-55921-4_4

An important point to note is that policies by which to pursue economic growth do not lead to the resolution of environmental problems, even when the existence of the EKC is confirmed. The work of Arrow et al. [1], as mentioned in Sect. 2.5.1, should not be denied. As for global warming, it is inevitable that there will be a continuous increase in carbon dioxide (CO_2) emissions worldwide in line with economic growth, as most countries belong to the non-Annex I group—that is, they are in the process of economic development, and have a high turning-point level.

It is necessary to reduce the turning-point level of the EKC, if emissions worldwide are to be controlled. The following points should very much be considered.

First, policy-makers need to convince people that CO_2 is a pollutant that inflicts enormous damage on future conditions. At the same time, policy-makers need to convince people that environmental improvement is brought about through institutional reforms or the introduction of environmental regulations, in connection with an awareness of the related social costs. Although international global warming countermeasures (including the Kyoto Protocol) have been expected to play an important role, the Kyoto Protocol seems to have had no great effect on reducing the turning point of the EKC, as seen in Sect. 3.4. Thus, the introduction of environmental regulations other than the Kyoto Protocol—for instance, carbon taxes—is an important option. In addition, the role of frameworks and rules of international corporation after 2020 will be of utmost importance.

Second, on the left-hand side of the EKC, in which most of the world's countries find themselves—that is, the case of environmental degradation in accordance with an increase in income—emission reduction is difficult, since abatement costs are thought to be relatively higher than the related social costs. Under the present circumstances, effective measures by which to reduce CO_2 emissions have not been found, save for energy conservation. One of the political implications of the EKC is the provision of economic assistance to developing countries: to promote emission reduction among these countries, the widespread proliferation of energy conservation technologies will be initially urgent. CO_2 emission reduction technologies, in particular, should be developed urgently. The development and introduction of environmental technologies, technology transfers, and development assistance are expected to shift the EKC of each country to the lower left.

4.2 Further Research

The following subjects are not sufficiently examined in this Brief, and should be addressed in future research.

The first subject is estimation modeling and its elaboration. Although the cubic polynomial of income was used in this analysis—in line with previous research—it seems to be essential to examine estimation models other than the polynomial model of income. A nonlinear function model may be more appropriate than the polynomial model. Building a structural model based on the theoretical model is highly desirable. Reviewing explanatory variables other than income may be insufficient, and the use of random coefficient models is worthy of consideration, if EKC estimation models featuring homogeneous parameters for each individual are inadequate.

The second subject is the interpretation of estimation results. Although an N-shaped EKC is estimated by the dynamic panel data model used here, interpretation is difficult. The turning point of the whole world sample is approximately USD80,000, and it is thought that it will be quite some time, if ever, before all countries worldwide will reach this income level. Therefore, one may be able to understand that the EKC is substantively an inverted U-shape. In any case, a society with increased CO_2 emissions and an increase in income cannot be said to be sustainable; for this reason, it is essential that social institutions be built so as not to increase CO_2 emissions, and it is equally essential that EKC research be deepened.

The third subject is the effects of CO_2 on the environment. As shown in Sect. 2.4, in the long term and while using the theoretical model, differences between flow-type and stock-type pollutants are thought to have no impact on the rate of economic growth. In reality, however, the negative impact of CO_2 stock is considered enormous.[1] Both theoretical and empirical research is needed to address this problem.

The fourth subject is the factors by which the EKC is shifted. Although the author briefly examined in Sect. 3.4 the relationship between the EKC and the Kyoto Protocol, this review seems to be insufficient and should be deepened. Moreover, it is interesting to investigate the relationship between the EKC shift and technological progress. The problem is considered important and worthy of investigation—even though some time will first need to pass, before a sufficient volume of data is available.

These subjects remain for future research.

4.3 Toward the Stationary State of John Stuart Mill

One cannot overemphasize the importance of creating a society where the economy coexists with the environment—with regard not only to greenhouse gases, but all pollutants. John Stuart Mill describes the stationary state in his *Principles of Political Economy* (1848), as follows.[2]

> It is scarcely necessary to remark that a stationary condition of capital and population implies no stationary state of human improvement. There would be as much scope as ever for all kinds of mental culture, and moral and social progress; as much room for improving the Art of Living, and much more likelihood of its being improved, when minds ceased to be engrossed by the art of getting on.

---

[1] The Intergovernmental Panel on Climate Change (IPCC) speaks of the concept of "carbon budget" in the fifth assessment report; it describes that the cumulative amount of CO_2 in the atmosphere is critical factor contributing to an increase in future temperatures (IPCC [2]).

[2] See Mill [3]; Chapter VI, "Of the Stationary State", in Book IV, "Influence of the Progress of Society on Production and Distribution".

As we see, Mill's stationary state is a situation in which spiritual and cultural activities, as well as moral and social improvements, are all carried out continuously, even in the absence of a zero increase in physical capital stock and in population. Environmental and ecological systems also continuously improve. Although macroeconomic valuables such as production and consumption remain constant over time, qualitative improvement does continue. At this time, production is carried out only to replace capital stock and the like. In this context, we derive a picture of a stable society that differs completely from one where there is merely zero growth.

Mill seems to consider this stationary state and qualitative development as the essence of classical economics, and evaluates them highly as components of a socially desirable situation. On the other hand, neoclassical economics—the mainstream economics to which we adhere today—principally focus on quantitative growth, and consider it a problem of optimal resource allocation, over time, of products and production factors such as labor and capital. Consequently, neoclassical economics discuss the level of growth rate with respect to the steady state. This contrasts starkly with Mill's discussion. According to Mill, the presence of sustainable development— in which the environment coexists with the economy—implies development without growth, or an economic society where there is qualitative improvement but no quantitative increase.

References

1. Arrow, K. J., Bolin, B., Costanza, R., Dasgupta, P., Folke, C., Holling, C. S., et al. (1995). Economic growth, carrying capacity and the environment. *Science, 268*, 520–521. Reprinted in *Ecological Economics, 15*, 91–95, in *Environment and Development Economics, 1*, 104–110, and in *Ecological Applications, 6*, 13–15.
2. IPCC (2013). *Climate change 2013: The physical science basis (summary for policymakers)*, Intergovernmental panel on climate change. http://www.climatechange2013.org/images/report/ WG1AR5_SPM_FINAL.pdf.
3. Mill, J. S. (1848). *Principles of political economy with some of their applications to social philosophy*. London: John W. Parker, West Strand.

Appendix A
Stokey Model with Accumulative Pollutant

A.1 Derivation of Optimal Growth Path

In this section, the author derives the optimal growth path from the model in Sect. 2.4. Again, the problem that should be solved is as follows:

$$\max \quad \int_0^\infty e^{-\rho t} \left(\frac{c(t)^{1-\sigma} - 1}{1 - \sigma} - \frac{B}{\gamma} X(t)^\gamma \right) dt$$

$$\text{subject to} \quad \dot{k}(t) = A e^{gt} k(t)^\alpha z(t) - \delta k(t) - c(t)$$

$$\dot{X}(t) = A e^{gt} k(t)^\alpha L z(t)^\beta - \eta X(t)$$

$$k(0) = k_0, \quad X(0) = X_0.$$

The current-value Hamiltonian is set to

$$\mathcal{H} \equiv \frac{c(t)^{1-\sigma} - 1}{1 - \sigma} - \frac{B}{\gamma} X(t)^\gamma + \lambda(t)(A e^{gt} k(t)^\alpha z(t) - \delta k(t) - c(t))$$
$$- \mu(t)(A e^{gt} k(t)^\alpha L z(t)^\beta - \eta X(t)).$$

The shadow price of pollution stock $X(t)$ that brings about disutility becomes a negative value. Hence, the author thinks of $\mu(t) > 0$ by considering $-\mu(t)$ the shadow price.

From the maximum principle, the necessary conditions for optimization (first-order conditions) and transversality conditions are given as follows:

$$\frac{\partial \mathcal{H}}{\partial c(t)} = 0 \qquad (A.1)$$

$$\frac{\partial \mathcal{H}}{\partial z(t)} \geqq 0 \qquad (A.2)$$

© Development Bank of Japan 2016
K. Uchiyama, *Environmental Kuznets Curve Hypothesis and Carbon Dioxide Emissions*, Development Bank of Japan Research Series,
DOI 10.1007/978-4-431-55921-4

$$\frac{\partial \mathcal{H}}{\partial k(t)} = \rho\lambda(t) - \dot{\lambda}(t) \tag{A.3}$$

$$\frac{\partial \mathcal{H}}{\partial X(t)} = -\rho\mu(t) + \dot{\mu}(t) \tag{A.4}$$

$$\lim_{t\to\infty} e^{-\rho t}\lambda(t)k(t) = 0 \tag{A.5}$$

$$\lim_{t\to\infty} e^{-\rho t}\mu(t)X(t) = 0. \tag{A.6}$$

Note that $z(t) \in [0, 1]$, Eq. (A.2) has equality if $z(t)$ is an interior solution. Since $\mathcal{H}$ is a concave function with regard to control variables $c(t)$, $z(t)$, and state variables $k(t)$, $X(t)$, Eqs. (A.1)–(A.6) comprise the necessary and sufficient conditions for optimization.

Arranging Eqs. (A.1)–(A.4) with a consideration of $z(t) \in [0, 1]$, the author obtains Eqs. (2.9)–(2.12).

A.2   Growth Rate in the Steady State

This section examines the growth rate in the steady state, in which all the variables grow at a constant rate; this satisfies the optimization conditions derived in Sect. A.1.

As the author obtains $\frac{\dot{c}(t)}{c(t)} = -\frac{1}{\sigma}\frac{\dot{\lambda}(t)}{\lambda(t)}$ from Eq. (2.9), $\frac{\dot{\lambda}(t)}{\lambda(t)}$ is constant if $\frac{\dot{c}(t)}{c(t)}$ is constant in the steady state. Therefore, from Eq. (2.11), $\frac{y(t)}{k(t)}$ is constant—that is, $g_c = g_k$.

Because $\dot{k}(t) = y(t) - \delta k(t) - c(t)$ from Eq. (2.8), the author has $g_k \equiv \frac{\dot{k}(t)}{k(t)} = \frac{y(t)}{k(t)} - \delta - \frac{c(t)}{k(t)}$. This is constant in the steady state, and so $\frac{c(t)}{k(t)}$ is also constant. This implies $g_c = g_k$. From the above, the author obtains

$$g_y = g_k = g_c, \tag{A.7}$$

and finds the growth rates of y, k, and c are all equal in the steady state.

The author derives

$$g_y = -\frac{1}{\sigma}g_\lambda \tag{A.8}$$

$$g_z = \frac{1}{\beta - 1}(g_\lambda - g_\mu) \tag{A.9}$$

from Eqs. (2.9) and (2.10), respectively.

Moreover, from Eq. (2.7),

$$gx = \frac{x(t)}{X(t)} - \eta$$

is obtained. As both gx and η are constant in the steady state, $\frac{x(t)}{X(t)}$ is also constant.
Hence, the author has

$$gx = g_X. \tag{A.10}$$

The author can write $x(t) = y(t)Lz(t)^{\beta-1}$ from Eq. (2.5); then, the author has

$$gx = g_y + (\beta - 1)g_z. \tag{A.11}$$

Equations (A.9)–(A.11) give rise to

$$gx = g_X = g_y + g_\lambda - g_\mu, \tag{A.12}$$

and therefore

$$(\gamma - 1)gx = g_\mu \tag{A.13}$$

is obtained from Eq. (2.12).

The author derives $gx = g_X = g_y - \sigma g_y - (\gamma - 1)gx$ from Eqs. (A.8), (A.12), and (A.13); through rearrangement, the author can obtain the following relation:

$$gx = g_X = \frac{1-\sigma}{\gamma} g_y. \tag{A.14}$$

This expresses the change rate of the pollution emissions in the stationary state.

From Eqs. (2.4), (A.7), (A.8), (A.13), and (A.14), the growth rate of y in the steady state can be written as

$$g_y = g + \alpha g_k + g_z = g + \alpha g_y + \frac{1}{\beta - 1}(g_\lambda - g_\mu)$$

$$= g + \alpha g_y - \frac{\sigma}{\beta - 1} g_y - \frac{1-\sigma}{\beta - 1}\frac{\gamma - 1}{\gamma} g_y;$$

by rearranging this the author obtains

$$g_y = \frac{\gamma(\beta - 1)g}{(1 - \alpha)\gamma(\beta - 1) + (\gamma + \sigma - 1)}. \tag{A.15}$$

Appendix B
Nonstationary Panel Data

B.1 Panel Unit Roots Tests

As panel unit root tests, the test of Levin et al. [7]—which is an extension of the test of Levin and Lin [6] and Im et al. [5]—the Fisher-type augmented Dickey–Fuller (ADF) test of Maddala and Wu [8] and Choi [2], and the test of Hadri [3], among others, have been proposed thus far.[1]

These tests are roughly classified into two types. To understand their features, we consider the following AR(1) process:

$$y_{it} = \alpha_i y_{i,t-1} + x'_{it} \beta_i + u_{it},$$

where i denotes cross-section and t expresses time. x_{it} is an exogenous variable used to express constant and trend terms.

One of the classification criteria is based on the assumption of the existence of homogeneous unit roots ($\alpha_1 = \alpha_2 = \cdots = \alpha_n = \alpha$). Another is based on the assumption that α_i differs from each other (heterogeneous).

The tests that apply the former assumption are:

- The Levin et al. [7] (LLC) test, which applies the ADF model and assumes the existence of homogeneous unit roots ($\alpha_1 = \alpha_2 = \cdots = \alpha_n = \alpha$) for each individual.
- The Hadri [3] test, which analyzes the residuals by ordinary least squares (OLS) regression, with either a constant term or both a constant and a trend term, and tests Lagrange multiplier (LM) statistics.

The tests that apply the latter assumption are:

---

[1] Baltagi [1] summarizes panel unit root tests with finesse.

© Development Bank of Japan 2016
K. Uchiyama, *Environmental Kuznets Curve Hypothesis and Carbon Dioxide Emissions*, Development Bank of Japan Research Series,
DOI 10.1007/978-4-431-55921-4

- The Im et al. [5] (IPS) test, which assumes heterogeneous α_i and is therefore equivalent to an individual unit root test. More specifically, it uses as test statistics the average statistical values of α_i obtained from an individual ADF test regression.
- The Fisher-type ADF (F-ADF) test, which tests the aggregation of individual ADF test statistics, as well as IPS. More specifically, it tests the combination of p-value of α_i obtained from individual ADF test regressions.

The null hypothesis of the Hadri test is that the panel contains no unit root; that of the other tests, meanwhile, is that the panel contains a unit root. More precisely,

$$
\begin{array}{lll}
\text{LLC} & H_0: \alpha = 1, & H_1: \alpha < 1 \\
\text{Hadri} & H_0: \alpha < 1, & H_1: \alpha = 1 \\
\text{IPS, F-ADF} & H_0: \alpha_i = 1 & \text{for all } i, \\
& H_1: \alpha_i < 1 & \text{for at least one } i,
\end{array}
$$

where H_0 denotes the null hypothesis, and H_1 an alternative hypothesis.

B.2   Panel Cointegration Tests

As panel cointegration tests, several tests are proposed, including the tests of Mc-Coskey and Kao [9] and Pedroni [11], as well as those of Kao [4] and Pedroni [10], which are employed in Sect. 3.2.2.[2] Outlines of the Kao [4] and Pedroni [10] tests are as follows.

Kao [4] test assumes the same autoregressive coefficients among all individuals. A DF test and an ADF test are applied for the residuals of panel estimation. The null hypothesis is given as $H_0: \rho = 1$ (no cointegration), while the alternative hypothesis is $H_1: \rho < 1$, where ρ expresses an autoregressive coefficient obtained from the residuals of OLS estimation. There are five test statistics, of which one test statistic is based on ADF test statistics and four on DF test statistics.

- DF-ρ, DF-t assume strict exogeneity with no correlation between regressors and disturbances.
- DF-ρ^*, DF-t^* consider the endogeneity of the regressors.

All the test statistics take an asymptotically normal distribution. The test is a one-sided test, and the smaller the test statistics are (including negative values), the greater the possibility of rejecting the null hypothesis.

Pedroni [10] test has a feature that allows for different autoregressive coefficients among individuals in some test statistics. There are seven test statistics, of which four (panel ν, panel ρ, panel t (nonparametric), and panel t (parametric)) are classified as within-type (panel type) statistics, which assume homogeneous autoregressive coefficients of residuals (i.e., $\rho_i = \rho$ (for all i)). The null hypothesis is $H_0: \rho_i =$

---

[2]Baltagi [1] gives a short survey of panel cointegration tests.

$\rho = 1$ (for all i), and the alternative hypothesis is H_0: $\rho_i = \rho < 1$ (for all i). The other three (group ρ, group t (nonparametric), and group t (parametric)) are classified as between-type (group type) statistics, which do not assume homogeneous autoregressive coefficients of residuals. The null hypothesis is H_0: $\rho_i = 1$ (for all i), and the alternative hypothesis is H_0: $\rho_i < 1$ (for all i). The test statistics of panel ν, panel/group ρ, and panel/group t (nonparametric) are based on nonparametric methods—that is, they adopt a procedure resembling the Phillips–Perron test, as a measure by which to control autocorrelation, while panel/group t (parametric) statistics employ parametric methods.

The following are short outlines of the McCoskey and Kao [9] and Pedroni [11] tests.

McCoskey and Kao [9] test has the null hypothesis of the existence of cointegration, in contrast to other tests. The test is an LM test that is based on the residuals of panel estimation. It utilizes the fully modified OLS (FMOLS) of Phillips and Hansen [12], among other factors, when estimating the parameters of the cointegration vector and analyzing the residuals.

Pedroni [10] test is a nonparametric test that resembles the Phillips–Perron test, which is a unit root test. It has a weak assumption regarding the distribution of disturbances, compared to the parametric DF test and ADF test. It also assumes strict exogeneity among the regressors.

References

1. Baltagi, B. H. (2013). *Econometric analysis of panel data* (5th ed.). West Sussex: Wiley.
2. Choi, I. (2001). Unit root tests for panel data. *Journal of International Money and Finance*, 20, 249–272.
3. Hadri, K. (2000). Testing for stationarity in heterogeneous panel data. *Econometric Journal*, 3, 148–161.
4. Kao, C. (1999). Spurious regression and residual-based tests for cointegration in panel data. *Journal of Econometrics*, 90, 1–44.
5. Im, K. S., Pesaran, M. H., & Shin, Y. (2003). Testing for unit roots in heterogeneous panels. *Journal of Econometrics*, 115, 53–74.
6. Levin, A., & Lin, C.-F. (1993). Unit root tests in panel data: New results. Discussion Paper No. 93–56, Department of Economics, University of California at San Diego.
7. Levin, A., Lin, C.-F., & Chu, C. S. J. (2002). Unit root tests in panel data: Asymptotic and finite-sample properties. *Journal of Econometrics*, 108, 1–24.
8. Maddala, G. S., & Wu, S. (1999). A comparative study of unit root tests with panel data and a new simple test. *Oxford Bulletin of Economics and Statistics*, 61, 631–652.
9. McCoskey, S., & Kao, C. (1998). A residual-based test of the null of cointegration in panel data. *Econometric Reviews*, 17, 57–84.
10. Pedroni, P. (1999). Critical values for cointegration tests in heterogeneous panels with multiple regressors. *Oxford Bulletin of Economics and Statistics*, 61, 653–678.
11. Pedroni, P. (2004). Panel cointegration: Asymptotic and finite sample properties of pooled time series tests with an application to the PPP hypothesis. *Econometric Theory*, 20, 597–625.
12. Phillips, P. C. B., & Hansen, B. E. (1990). Statistical inference in instrumental variables regression with I(1) processes. *Review of Economic Studies*, 57, 99–125.

Appendix C
Model Specification Test

This section discusses the model specification of the Arellano and Bond [1] dynamic panel data model used in Sect. 3.3.3.

One of the important assumptions with regards to the dynamic panel data model is that disturbance u_{it} has no autocorrelation of the second-order or more. If this assumption is not satisfied, instrumental variables become inappropriate. Any generalized method of moments (GMM) estimate obtained by using such instrumental variables will not have consistency; therefore, Arellano and Bond [1] propose a test for autocorrelation (i.e., the Arellano–Bond test). This tests the existence of first-order and second-order autocorrelations with regard to disturbances. The null hypothesis is that there is no autocorrelation, and the test statistics follow an asymptotically standard normal distribution.[3]

Arellano and Bond [1] also propose a test for over-identifying restrictions for instrumental variables (i.e., the Sargan test).[4]

The null hypothesis of the test is that the over-identifying restrictions are valid. The following two points have been highlighted as test limitations. First, the power is low for a small sample. Second, the test has the tendency to reject the null hypothesis excessively when the disturbance has heteroskedasticity.

Reference

1. Arellano, M., & Bond, S. (1991). Some tests of specification for panel data: Monte Carlo evidence and an application to employment equations. *Review of Economic Studies, 58*, 277–297.

---

[3] See Arellano and Bond [1] for the test statistics.

[4] The test follows the same thinking as Hansen's J test.

© Development Bank of Japan 2016
K. Uchiyama, *Environmental Kuznets Curve Hypothesis and Carbon Dioxide Emissions*, Development Bank of Japan Research Series,
DOI 10.1007/978-4-431-55921-4

Further Reading

1. Abdullah, S., & Morley, B. (2014). Environmental taxes and economic growth: Evidence from panel causality tests. *Energy Economics*, *42*, 27–33.
2. Aghion, P., & Howitt, P. (1998). *Endogenous growth theory*. Cambridge: MIT Press.
3. Arellano, M. (2003). *Panel data econometrics*. Oxford: Oxford University Press.
4. Arellano, M., & Bover, O. (1995). Another look at the instrumental variables estimation of error-components models. *Journal of Econometrics*, *68*, 29–51.
5. Auffhammer, M., & Carson, R. T. (2008). Forecasting the path of China's CO_2 emissions using province-level information. *Journal of Environmental Economics and Management*, *55*, 229–247.
6. Baek, J. (2015). Environmental Kuznets curve for CO_2 emissions: The case of Arctic countries. *Energy Economics*, *50*, 13–17.
7. Barro, R. J., & Sala-i-Martin, X. (2003). *Economic growth* (2nd ed.). Cambridge: MIT Press.
8. Bernard, J. T., Gavin, M., Khalaf, L., & Voia, M. (2015). Environmental Kuznets curve: Tipping points, uncertainty and weak identification. *Environmental and Resource Economics*, *60*, 285–315.
9. Blundell, R., & Bond, S. (1998). Initial conditions and moment restrictions in dynamic panel data models. *Journal of Econometrics*, *87*, 115–143.
10. Culas, R. J. (2007). Deforestation and the environmental Kuznets curve: An institutional perspective. *Ecological Economics*, *61*, 429–437.
11. de Bruyn, S. M. (1997). Explaining the environmental Kuznets curve: Structural change and international agreements in reducing sulphur emissions. *Environment and Development Economics*, *2*, 485–503.
12. de Bruyn, S. M., van den Bergh, J. C. J. M., & Opschoor, J. B. (1998). Economic growth and emissions: Reconsidering the empirical basis of environmental Kuznets curves. *Ecological Economics*, *25*, 161–175.
13. Ekins, P. (1997). The Kuznets curve for the environment and economic growth: Examining the evidence. *Environment and Planning*, *29*, 805–830.

© Development Bank of Japan 2016
K. Uchiyama, *Environmental Kuznets Curve Hypothesis and Carbon Dioxide Emissions*, Development Bank of Japan Research Series,
DOI 10.1007/978-4-431-55921-4

14. Grossman, G. M. (1995). Pollution and growth: What do we know? In I. Goldin & L. A. Winters (Eds.), *The economics of sustainable development*. Cambridge: Cambridge University Press.
15. He, J., & Richard, P. (2010). Environmental Kuznets curve for CO_2 in Canada. *Ecological Economics, 69*, 1083–1093.
16. Hettige, H., Lucas, R. E. B., & Wheeler, D. (1992). The toxic intensity of industrial production: Global patterns, trends, and trade policy. *American Economic Review, 82*, 478–481.
17. Hsiao, C. (2014). *Analysis of panel data* (3rd ed.). Cambridge: Cambridge University Press.
18. Huang, W. M., Lee, G. W. M., & Wu, C. C. (2008). GHG emissions, GDP growth and the Kyoto protocol: A revisit of environmental Kuznets curve hypothesis. *Energy Policy, 36*, 239–247.
19. Islam, N., Vincent, J., & Panayotou, T. (1998). Unveiling the income-environment relationship: An exploration into the determinants of environmental quality. Working Paper, 98–12, Emory University.
20. John, A., Pecchenino, R., Schimmelpfennig, D., & Schreft, S. (1995). Short-lived agents and the long-lived environment. *Journal of Public Economics, 58*, 127–141.
21. Jones, L. E., & Manuelli, R. E. (2001). Endogenous policy choice: The case of pollution and growth. *Review of Economic Dynamics, 4*, 369–405.
22. Kahn, M. E. (1998). A household level environmental Kuznets curve. *Economics Letters, 59*, 269–273.
23. Kao, C., & Chiang, M. H. (2000). On the estimation and inference of a cointegrated regression in panel data. *Advances in Econometrics, 15*, 179–222.
24. Kuznets, S. (1965). *Economic growth and structure: Selected essays*. New York: W. W. Norton & Co.
25. Kuznets, S. (1966). *Modern economic growth: Rate, structure, and spread*. New Haven: Yale University Press.
26. Levin, A., & Lin, C.-F. (1992). Unit root tests in panel data: Asymptotic and finite-sample properties. Discussion Paper No. 92–93, Department of Economics, University of California at San Diego.
27. Lindmark, M. (2002). An EKC-pattern in historical perspective: Carbon dioxide emissions, technology, fuel prices and growth in Sweden 1870–1997. *Ecological Economics, 42*, 333–347.
28. Lipford, J. D., & Yandel, B. (2010). Environmental Kuznets curve, carbon emissions, and public choice. *Environment and Development Economics, 15*, 417–438.
29. Lucas, R. E. B., Wheeler, D., & Hettige, H. (1992). Economic development, environmental regulation and the international migration of toxic industrial pollution: 1960–1988. In P. Low (Eds.), *International trade and the environment*, World Bank Discussion Paper No. 159, World Bank.
30. Managi, S. (2006). Are there increasing returns to pollution abatement? Empirical analytics of the environmental Kuznets curve in pesticides. *Ecological Economics, 58*, 617–636.

31. Munasinghe, M. (1999). Is environmental degradation an inevitable consequence of economic growth: Tunneling through the environmental Kuznets curve. *Ecological Economics, 29,* 89–109.
32. Panayotou, T. (1993). Empirical tests and policy analysis of environmental degradation at different stages of economic development. Working Paper, WP238, Technology and Employment Programme, International Labor Office.
33. Panayotou, T. (1997). Demystifying the environmental Kuznets curve: Turning a black box into a policy tool. *Environment and Development Economics, 2,* 465–484.
34. Phillips, P. C. B., & Hansen, B. E. (1990). Statistical inference in instrumental variables regression with I(1) processes. *Review of Economic Studies, 57,* 99–125.
35. Stern, D. I., Common, M. S., & Barbier, E. B. (1996). Economic growth and environmental degradation: The environmental Kuznets curve and sustainable development. *World Development, 24,* 1151–1160.
36. Uzawa, H. (1965). Optimal technical change in an aggregative model of economic growth. *International Economic Review, 6,* 18–31.
37. Uzawa, H. (2003). *Economic theory and global warming.* Cambridge: Cambridge University Press.
38. Westerlund, J., & Basher, S. A. (2008). Testing for convergence in carbon dioxide emissions using a century of panel data. *Environmental and Resource Economics, 40,* 109–120.
39. Wooldridge, J. M. (2010). *Econometric analysis of cross section and panel data* (2nd ed.). Cambridge: MIT Press.
40. Yamaguchi, Y., Sonobe, T., & Otsuka, K. (2007). Beyond the environmental Kuznets curve: A comparative study of SO_2 and CO_2 emissions between Japan and China. *Environment and Development Economics, 12,* 445–470.